Edward Whymper

How to Use the Aneroid Barometer

Edward Whymper

How to Use the Aneroid Barometer

ISBN/EAN: 9783744764322

Printed in Europe, USA, Canada, Australia, Japan

Cover: Foto ©berggeist007 / pixelio.de

More available books at **www.hansebooks.com**

HOW TO USE

THE

ANEROID BAROMETER

BY

EDWARD / WHYMPER

I. COMPARISONS IN THE FIELD
II. EXPERIMENTS IN THE WORKSHOP
III. UPON THE USE OF THE ANEROID BAROMETER
IN DETERMINATION OF ALTITUDES
IV. RECAPITULATION

NEW YORK

CHARLES SCRIBNER'S SONS, 743, 745 BROADWAY

1891

ZOOLOGICAL COLLECTIONS
FROM THE GREAT ANDES OF THE EQUATOR.

WALTER BURTON, Naturalist, of 191 WARDOUR ST., LONDON, W., begs to inform Collectors of BIRDS, INSECTS, REPTILES, and FISHES that he has the collections formed by EDWARD WHYMPER, Esq., upon his Journey in Ecuador for sale, including :—

The first set of the DIURNAL LEPIDOPTERA, classified in 14 boxes, embracing 240 specimens of 106 Species, taken at various elevations up to 16,000 feet above the sea, with Localities and Descriptions attached ; duplicates of *Colias dimera*, Doubl. and Hew., *Pieris xanthodice*, Lucas, *Pieris elodia*, Boisd., etc., 1s. 6d. to 2s. each ; numerous skins of Humming and other BIRDS, in the very best condition ; excellent specimens of the very rare and interesting FISH, *Cyclopium cyclopum*, Humboldt (see *Travels amongst the Great Andes*, chap. xii, and the description in *Supp. App.* by the late Dr. F. Day), 4s. to 7s. each ; amongst the FROGS perfect examples of males, females and young of the curious *Nototrema marsupiatum* (Dum. & Bibr.) ; also of the new species *Hylodes Whymperi*, described by Mr. G. A. Boulenger, captured from 11,000 to 13,200 feet ; *Hylodes conspicillatus*, Gthr. ; *H. unistrigatus*, Gthr., and young ; *Phryniscus lævis*, Gthr. ; *Dendrobates tinctorius* (Schneid.) ; *Bufo agua*, etc. etc., 3s. to 7s. 6d. each. Fine specimens of the LIZARDS *Liocephalus trachycephalus*, *L. iridescens*, Gthr. ; *Ecpleopus* (*Pholidobolus*) *montium*, Peters ; etc. etc., 5s. to 7s. each. Examples of the SNAKES *Liophis reginæ*, L., var. *albiventris*, Jan. ; *L. splendens*, Jan. ; *Bothrops atrox*, (L.) ; *B. Schlegeli*, (Berthold) ; *Leptognathus nebulatus*, (L.) ; *Elaps Marcgravi*, (Wied.), var. *ancolaris*, Jan. ; *Oxyrhopus clælia*, Jan. ; *Boa constrictor*, L., etc. etc., 5s. to 7s. each. Specimens of the SCORPIONS *Broteas subnitens*, Gervais (from 10,000 feet and upwards), and *Centrurus Americanus*, De Geer ; of the AMPHIPOD *Hyalella inermis*, S. I. Smith (obtained from 12,000 to 13,000 feet) ; and the fine BUG *Carineta basalis*, Walker. Good series of the new species of ANT *Pheidole monticola* (females, soldiers and workers) described by Mr. P. Cameron ; and specimens of *Camponotus atriceps*, Sm., *C. sylvaticus*, Oliv., *Atta sexdens*, L., *Pseudomyrma gracilis*, Fabr., *Ectatomma quadridens*, Fab., and *Pachycondyla carbonaria*, Smith, etc. etc., 1s. to 1s. 6d. each. And a numerous series of the special COLEOPTERA of the highest zones of the Ecuadorian Andes, including the following New Species which are described in the *Supplementary Appendix* to *Travels amongst the Great Andes of the Equator :—Leucopelœa albescens*, Bates ; *Colpodes steno*, Bates ; *C. alticola*, Bates ; *Pterostichus integer*, Bates ; *Anisotarsus bradytoides*, Bates ; *Pelmatellus Andium*, Bates ; *P. variipes*, Bates ; *P. oxynodes*, Bates ; *Astylus bis-sexguttatus*, Gorham ; *Ananca debilis*, Sharp ; *Meloe sexguttatus*, Sharp ; *Sterculia impressipennis*, Sharp ; *Epitragus dilutus*, Sharp ; *Philonthus Whymperi*, Sharp ; *Silpha microps*, Sharp, etc. etc., 1s. 6d. to 2s. 6d. each.

WALTER BURTON, 191 WARDOUR STREET, LONDON.

HOW TO USE

THE

ANEROID BAROMETER

BY

EDWARD WHYMPER

I. COMPARISONS IN THE FIELD
II. EXPERIMENTS IN THE WORKSHOP
III. UPON THE USE OF THE ANEROID BAROMETER
IN DETERMINATION OF ALTITUDES
IV. RECAPITULATION

NEW YORK

CHARLES SCRIBNER'S SONS, 743, 745 BROADWAY

1891

PREFACE.

THE following paper is divided into four sections. The first records comparisons of the aneroid against the mercurial barometer in the field ; the second is concerned with experimental research in the workshop ; the third is occupied by practical considerations arising from the facts recorded ; and the fourth is composed of a recapitulation of the principal points which are dwelt upon in the previous parts, and forms an index to the whole.

The investigations which are recorded in these pages have extended over eleven years. In 1879–80, having occasion to pass some length of time at great elevations in the Equatorial regions of South America, I took the opportunity to make comparisons of the aneroid against the mercurial barometer at low pressures. Some of these comparisons are given in the First Part.[1]

On entering upon this journey, I entertained the hope that close approximations to true atmospheric pressure might be obtained by employing several or a number of aneroids. The idea was that the plus errors of some instruments might balance the minus errors of others, and that means of the whole might come near the truth. This hope was speedily dissipated. Each individual instrument indicated lower pressures than the mercurial barometer, and means of the whole were, consequently, less than the truth. Deduced altitudes were much too high. In strong contrast to these unwelcome facts there were others equally perplexing, although less unpleasant. It was found that even when the aneroids had acquired minus errors of *one to two inches* they might on differences of level of several thousand feet indicate nearly the same *differences of pressure* as the mercurial barometer. Ascending

---

[1] Part 1 is reprinted from the Appendix to *Travels amongst the Great Andes of the Equator*. This volume contains an account of the journey, and describes the circumstances under which the comparisons were made. These two works are published simultaneously, and can be obtained separately.

B

observations of aneroids, it was found, never agreed with descending ones, and the latter always indicated less difference of pressure and less difference of level than the former, if they were made only a short time after them ; but, in such cases, means of the ascending and descending observations closely accorded with means of similar observations made with the mercurial barometer.

A long time elapsed after my return from this journey before I commenced to find a solution of these puzzles. I examined and rearranged my observations for several years before it occurred to me to tabulate the means of the whole of the aneroid readings in chronological order, to do the same for the readings of each individual instrument, and to take out the differences from the mercurial barometer of each mean and of each individual reading. When this was done I saw that the results were influenced by the lapse of time between the readings as well as by variations in pressure. The general conclusions at which I then arrived are stated upon pp. 9, 14.

Having introduced the matter to the notice of Mr. J. J. Hicks, of 8 Hatton Garden, and gained his co-operation, I then proceeded to test aneroids under varied conditions of time and pressure ; in the end finding explanations for the points of perplexity, and that aneroids, whilst largely differing amongst each other, all follow rules which have the force of law.

The prosecution of these experiments has extended over a far greater length of time than appeared necessary at their inception, and has absorbed almost all my leisure during the last few years. The labour will have been well bestowed, and I shall attain my aim, if in these pages I succeed in explaining some points in the behaviour of aneroids which have undoubtedly been puzzling, and have seemed contradictory and anomalous ; and thus render it possible for the ever-increasing number of those who employ these instruments in the field to use them with more confidence and certainty.

EDWARD WHYMPER.

PART I.
COMPARISONS OF THE ANEROID AGAINST THE MERCURIAL BAROMETER.

Aneroids were carried to Ecuador to endeavour to ascertain whether the *means* of the readings of several, or of a number, would or would not accord with the mercurial barometer at low pressures.

It has long been known that the indications afforded by a *single* aneroid are apt to be of a very deceiving nature, even at moderate elevations ; but it seemed to me *possible* if several, or if a number were employed, that one might, by inter-comparison, discriminate between those which went most astray and those which held closely together ; and that, by adoption of the means of the readings of the latter, a decent approximation might be obtained to the truth, possibly even at great altitudes. It may be added that I wished this might prove to be the case ; for the portability of aneroids, the facility with which they can be read, and the quickness of their action, would render them valuable for many purposes, if their indications could be relied upon.

Eight aneroids were taken. One of these, by Casella, marked No. 580, had been made for an earlier journey ; and, through being only graduated to 20 inches, was of no service for comparison at great heights.[1] The seven remaining aneroids were constructed expressly for the expedition, and were under trial and examination for nearly twelve months before our departure. They were selected from picked instruments, and only those were taken such as were, so far as one could tell, in all respects as perfect as could be produced.[2] These seven aneroids were marked A – G. A, B, and C were graduated from 31 down to 15 inches, and D, E, F, G were graduated from 31 down to 13 inches,—a range which I thought would be sufficient for my purposes.[3]

It became apparent at an early stage of the journey, *a.* that the whole of the aneroids had acquired considerable errors ; *b.* that they differed amongst each other to a very large extent ; and *c.* that neither means of the whole, nor means of those which held closest together, nor means of any combination, would give decent approximations to the truth. The more evident this became the greater importance I attached to the preservation of the mercurials. Comparisons of the aneroids against the mercurials were neverthe-

---

[1] This was left with Mr. Chambers at Guayaquil, as a reserve for him, in case accident befell the Standard Mercurial ; and he read both instruments during the whole of our absence in the interior.

[2] I abstain, however, from mentioning the names of the makers (to whom I am much indebted), lest the remarks which follow should be construed to their disadvantage.

[3] It proved to be inadequate.

less continued until the end of the journey;[1] and after two months' experience in the interior the behaviour of the aneroids in ascending and descending was so well ascertained that one might, I think, have made use of their indications to determine differences of level without committing very great mistakes.

In the following pages, I propose first to give some of my experiences, and then to draw such conclusions as appear to be warranted.[2]

§ 1. Shortly before my departure from London, I made (on October 25, 1879) a final comparison of the aneroids against the mercurial barometer. Only one of the aneroids corresponded exactly, and of the others some were too high and some were too low, the greatest difference between them amounting to 0·225 of an inch, and the mean of the whole showing an error of + 0·148 of an inch.

§ 2. Our ship stopped a clear day (November 20, 1879) at Jamaica, and I took the opportunity to carry the aneroids to the top of the Blue Mountains,[3] comparing them against the mercurial before starting and upon return, and comparing them against each other at the highest point attained. The following is the complete record, and it will be seen from it that the mean

No. of Barometer.	At start.	At top.	On return.
Aneroid 580 . . .	29·980	25·430	29·975
,, A . . .	29·850	25·500	29·850
,, B . . .	29·800	25·250	29·800
,, C . . .	29·700	25·120	29·650
,, D . . .	29·850	25·400	29·880
,, E . . .	29·800	25·300	29·750
,, F . . .	29·750	25·350	29·700
,, G . . .	29·800	25·310	29·700
Extreme differences .	0·280	0·380	0·325
Mean of aneroids . .	29·816	25·332	29·788
Merc. bar. No. 554 . .	29·876	. . .	29·854
Mean error of aneroids	− 0·060	. . .	− 0·066

error of the aneroids (which in London was + 0·148) had changed on arrival at Kingston to − 0·060, and upon return in the evening of November 20 it was still further increased to − 0·066.

[1] In all the comparisons which are made throughout this paper the readings of the mercurial barometer are reduced to 32° Faht.

[2] Paragraphs 1–10 should be read in connection with the tables at pp. 10, 11.

[3] Drove to Gordonstown, walked thence to Newcastle, and on until we came to a notch in the mountains commanding a view of the northern side of the island. Read the aneroids at this place.

§ 3. Comparisons were again made at Colon and Guayaquil,[1] and at neither of these places (at the level of the sea) was there any further increase in the mean error of the aneroids; but at Muñapamba (where we commenced to mount the outer Andean ranges) their difference amongst each other had risen to 0·500 of an inch, and the mean error was increased to − 0·098 of an inch.

§ 4. At Tambo Gobierno (the culminating point of the road over the outer Andean range), 10,417 feet above the sea, the mean error of the aneroids had risen to − 0·487 of an inch, and the extreme difference of their readings amounted to 0·715.

§ 5. With the descent on the other side the aneroids came more closely together, but their mean error continued to augment,—being upon arrival at Guaranda [2] (8894 feet) − 0·520, and it rose in one week to − 0·655. The 'greatest difference' also continued to increase, and it stood on Christmas Day at 0·800 of an inch. The following record, showing the continual increases in the errors, will be found interesting.

No. of Barometer.	Dec. 18, 1879.	Dec. 20, 1879.	Dec. 23, 1879.	Dec. 25, 1879.
Aneroid A	21·700	21·700	21·700	21·600
„ B	21·170	20·960	20·940	20·870
„ D	21·460	21·430	21·450	21·390
„ E	21·500	21·500	21·500	21·440
„ F	21·220	21·030	20·950	20·800
„ G	21·400	21·380	21·300	21·300
Extreme differences .	0·530	0·740	0·760	0·800
Mean of aneroids . .	21·408	21·333	21·321	21·233
Mean Merc. bar. . .	21·928	21·912	21·934	21·888
Mean errors of aneroids	− 0·520	− 0·579	− 0·613	− 0·655

§ 6. Upon December 26, 1879, we encamped on the Arenal (14,375 feet), at the foot of Chimborazo, and on the morning of the 27th the mean error of the aneroids was found to be − 0·737, and their greatest difference 0·880.

§ 7. We then moved up to the Second Camp on Chimborazo (16,664 feet), the mean error continuing to rise, and amounting upon December 30 to − 0·903.

§ 8. Upon arrival at the Third Camp (17,285 feet) I found that comparisons at greater heights would have to be made between five aneroids

---

[1] As aneroid 580 was left at Guayaquil, the comparisons are now between the seven remaining instruments.

[2] Aneroid C was lost or stolen shortly before arrival at Guaranda, thus reducing the number under comparison to six.

only, as the error which B had acquired was so large that we had already got *beyond its range.* The mean error of the aneroids at this point amounted to − 0·974, and their greatest difference to 1·120 inches.

§ 9. The aneroids D and E were alone taken to the summit of Chimborazo on the first ascent, January 4, 1880, and these two were taken because they were working better than the others. The readings on the summit are instructive.

Merc. bar. No. 558	14·110 inches.
Aneroid D	13·050 „
do. E	12·900 (by estimation).

The mean of the two aneroids is seen to be 12·975 inches, and the error of this upon the mercurial − 1·135 inches. I defer comment to a later point.

§ 10. Their prolonged residence upon Chimborazo seriously affected the constitutions of aneroids F and G. The index of the latter instrument became immovable, and the former was afflicted with a quivering action which set observation at defiance. Comparisons for the remainder of the journey were thus restricted to A, B, D, E only, and they are given in the tables upon pp. 10, 11 so far as is necessary to support the statements,— *a.* that the aneroids acquired considerable errors ; *b.* that they differed amongst each other to a very large extent ; and *c.* that their means were far from the truth.

§ 11. After we had been three to four weeks in the interior, the aneroids A, B, D, E were found to hold pretty constantly together (or, speaking more correctly, their movements were harmonious), and they seemed to have acquired their maximum errors for the pressures at which they were used.[1] Of the above four instruments B had the largest index-error, and the following table shows that it remained tolerably constant. It then became interesting

Date.	Place of observation.	Merc. bar. 558.	Aneroid B.	Error of B.
Jan. 14, 1880	Chuquipoquio .	19·683 inch	17·820 inch	− 1·863 inch
Feb. 8, „	Hac. de la Rosario	20·805 „	19·100 „	− 1·705 „
„ 9, „	Illiniza (S. side) .	17·239 „	15·400 „	− 1·839 „
„ 16, „	On Cotopaxi . .	17·431 „	15·650 „	− 1·781 „
„ 26, „	Machachi . . .	21·142 „	19·360 „	− 1·782 „
Mar. 28, „	Hac. Guachala .	21·618 „	19·950 „	− 1·668 „
May 15, „	Quito	21·631 „	19·990 „	− 1·641 „
June 8, „	Illiniza (N. side).	17·222 „	15·400 „	− 1·822 „

to observe whether aneroids which had acquired such large index-errors could be usefully employed for the determination of differences of level.

[1] See the last column of the table at p. 11.

§ 12. Upon the journey to the north of Quito I carried merc. bar. No. 558 and aneroids A and B, and upon arrival at the top of the great ravine of Guallabamba took simultaneous observations of the mercurial and the aneroids. At the bottom of the ravine, two hours and a half later, readings of all three were repeated with the following result:—

Date.	Barometer.	Read at top.	Read at bottom.
Mar. 27, 1880	Merc. bar. 558 (red. to 32° Faht.)	21·692	23·929
- do.	Aneroid A	21·140	23·400
do.	do. B	19·940	22·200

The rise of the Mercurial Barometer was 2·237 inches.
do.   do.   Aneroid A   „ 2·260 „
do.   do.   „ B   „ 2·260 „

§ 13. The foregoing experiment is a *descending* one employing two aneroids, and the next is an *ascending* one in which three were observed. Upon the occasion of the attempt to ascend Illiniza from the north, I read the three aneroids A, B, D before departure from Machachi (9839 feet), and did the same at our camp (15,446 feet).

Date.	Barometer.	At Machachi (9 a.m.)	At Camp (6 p.m.)
June 8, 1880	Merc. bar. 558	21·176 inches	17·222 inches.
do.	Aneroid A	20·650 „	16·810 „
do.	do. B	19·530 „	15·400 „
do.	do. D	20·290 „	16·380 „

The fall of the Mercurial Barometer was 3·954 inches.
The mean fall of the three aneroids was 3·960 „

§ 14. Upon the second ascent of Chimborazo I carried aneroids A and E to the summit, reading them at the fifth camp and at the top. Aneroid A became much out of range, and I therefore cannot give its reading.

Date.	Barometer.	Fifth Camp (4 a.m.)	Summit (2 p.m.)
July 3, 1880	Merc. bar. 558	16·931 inches	14·044 inches.
do.	Aneroid E	16·060 „	12·990 „

The fall of the Mercurial Barometer was 2·887 inches.
do.   Aneroid E   „ 3·070 „

§ 15. The examples which are quoted in §§ 12, 13, 14 give the closest coincidences that I can mention from amongst experiments of this order. Upon the whole, it appeared to me that better values could be obtained from aneroids by taking the *mean* of ascending and descending observations,[1] than by taking the means of either ascending or descending ones alone, and I now give an example in which this method of treatment was adopted.

On March 19, 1880, I carried the aneroids A, B, D, E from our lodging

[1] When ascent and descent are only a short space of time apart.

at Quito to the top of the hill called the Panecillo, on this occasion reading the 'scales of feet' upon them before departure, and again upon return to Quito. From the means of the ascending and descending readings, the summit of the Panecillo appears to be 651·25 feet above the level of the principal Plaza of Quito, which from the mean of twenty-two observations of mercurial barometer by myself is found to be 9343·3 feet above the sea. I have no observation of mercurial barometer on the Panecillo; and, if I had, should still quote by preference the independent observations of Messrs. Reiss and Stübel, who, from the mean of a large number of observations of mercurial barometer, give for the height of Quito 9350 feet, and for the Panecillo (two observations of m.b.) 10,007 feet. Their *difference of level* therefore is 657 feet, or 5 feet 9 inches more than the height indicated by the aneroids.

Barometer.	At Quito, in Hotel.	Summit of Panecillo.	= a rise of	On return to Quito.	= a fall of
Aneroid A	10,680 feet	11,325 feet	645 feet	10,760 feet	565 feet
do. B	12,310 „	13,050 „	740 „	12,390 „	660 „
do. D	11,260 „	11,950 „	690 „	11,340 „	610 „
do. E	11,000 „	11,680 „	680 „	11,060 „	620 „

Mean of ascending readings 688·75 feet.
Mean of descending do.   613·75 feet.
Mean of ascending and descending 651·25 feet.

§ 16. As the journey approached its termination, I became curious to observe how the aneroids would read against the mercurials upon return to the level of the sea. We arrived at Guayaquil again on July 13, 1880, and the barometers were compared against each other from the 16th to the 27th.[1] The error of aneroid A upon the 16th was − 0·361 of an inch, and of E − 0·321, but by the 27th their respective errors diminished to − 0·341 and − 0·291. I have not allowed the index of either to be altered. They continued to recover in the course of time; and I found, upon January 9, 1885, that aneroid E possessed an index-error of + 0·160, which was very nearly its error upon the last comparison in 1879 before our start, when it was seen to be + 0·182. Aneroid A did not recover with the same rapidity. Upon January 9, 1885, its error had diminished to − 0·200, and in five years more it recovered another tenth of an inch.

§ 17. In the tables at pp. 12, 13 the comparisons of the aneroids A and E are presented separated from the others. These two instruments were those which were most consistent in their behaviour, and were those which were most frequently employed. From inspection of the tables it will be immediately apparent that 'a good return' is of little value as a test of working.

---

[1] See the tables at pp. 12, 13 for this and for the succeeding paragraph.

Upon the last comparison before departure, these two aneroids possessed almost exactly similar index-errors (+ 0·172 and + 0·182), and upon return to Guayaquil their index-errors were not far apart (− 0·341 and − 0·291). It would have appeared legitimate to conclude that their working had closely corresponded, but inspection of the last two columns of the table shows that such a conclusion would have been extremely erroneous. The case of E, taken by itself, is still stronger. This, in course of time, ' returned ' almost perfectly ; and inasmuch as this instrument (like all the others) was tested before departure, inch by inch, against the mercurial barometer under the air-pump, and corresponded almost perfectly, it would have seemed right to conclude that its readings in the interim must have been nearly free from error. Yet this instrument, at the greatest height at which it was compared, was found to possess a minus error of an inch and a fifth, the value of which, at the elevation in question, exceeds *two thousand feet* (see § 9).

§ 18. Some of the more important conclusions which must be arrived at from consideration of the results of these comparisons of the aneroid against the mercurial barometer are so obvious that I consider it unnecessary even to point them out ; and, in the remarks which follow, I endeavour more to indicate the ways in which the aneroid may be advantageously used, than to emphasize the objections which might be urged against its employment.

A. It seems possible, without reference to a standard, by intercomparison of a number of aneroids, to discriminate between them, and to select those in which most confidence should be placed.

B. That, with aneroids of the present construction, it is unlikely that decent approximations to the truth will be obtained at low pressures, even when employing a large number of instruments. The errors of the whole series (A − G) were invariably minus ones, and in the worst cases amounted to as much as *two inches* upon the mercurial barometer.

C. That differences of level at great heights (low pressures) may be determined with considerable accuracy with aneroids, even when they have acquired very large index-errors.

D. That in observations of this description a nearer approach to the truth is generally obtained by employing the mean of ascending and descending readings than by taking ascending or descending readings separately.

E. That the test which is commonly applied of comparing for brief periods (minutes or hours) aneroids against mercurial barometers under the air-pump is of little or no value in determining the errors which will appear in aneroids used at low pressures for long periods (weeks or months).

F. That, similarly, comparisons of aneroids against mercurial barometers in balloon for a brief space of time afford little or no clue to the errors which will be exhibited by the former when subjected to low pressures for

No.	Date.	Place of Observation.	No. of Aneroids observed.	Greatest difference between Aneroids.	Mean of Aneroids.	Mercurial Barometer (red. to 32° Faht.)	Mean error of Aneroids.
1.	Oct. 25, 1879	London	8	0·225 inch.	29·916 inch.	29·768 inch.	+0·148 inch.
2.	Nov. 18, "	Off S. Domingo (at sea)	8	0·260 "	29·886 "	29·985 "	−0·099 "
3.	" 20, "	Kingston, Jamaica	8	0·280 "	29·816 "	29·876 "	−0·060 "
4.	" " "	Above Newcastle, Jamaica	8	0·380 "	25·332 "	...	...
5.	" " "	Kingston, do.	8	0·325 "	29·788 "	29·854 "	−0·066 "
6.	Dec. 2, "	Colon, Isthmus of Panama	8	0·360 "	29·790 "	29·845 "	−0·055 "
7.	" 10-12, "	Guayaquil, Ecuador	8	0·346 "	29·769 "	29·824 "	−0·055 "
8.	" 16, "	Muñapamba, do.	7	0·500 "	28·510 "	28·608 "	−0·098 "
9.	" 17, "	Tanbo Gobierno, Ecuador	7	0·715 "	20·272 "	20·759 "	−0·487 "
10.	" " "	San José de Chimbo, do.	7	0·660 "	21·955 "	...	...
11.	" 18, "	Guaranda, do.	6	0·530 "	21·408 "	21·928 "	−0·520 "
12.	" 20, "	Do. do.	6	0·740 "	21·333 "	21·912 "	−0·579 "
13.	" 23, "	Do. do.	6	0·760 "	21·321 "	21·934 "	−0·613 "
14.	" 25, "	Do. do.	6	0·800 "	21·233 "	21·888 "	−0·655 "
15.	" 27, "	Chimborazo, First Camp	6	0·880 "	17·135 "	17·872 "	−0·737 "
16.	" 28, "	Do. Second do.	6	0·780 "	15·643 "	16·476 "	−0·833 "
17.	" 29, "	Do. do. do.	6	0·830 "	15·611 "	16·488 "	−0·877 "

No.	Date.	Place of Observation.	No. of Aneroids observed.	Greatest difference between Aneroids.	Mean of Aneroids.	Mercurial Barometer (red. to 32° Faht.)	Mean error of Aneroids.
18.	Dec. 30, 1879	Chinborazo, Second Camp	6	0·825 inch.	15·577 inch.	16·480 inch.	-0·903 inch.
19.	Jan. 4-5, 1880	Do. Third do.	4	1·120 ,,	15·045 ,,	16·019 ,,	-0·974 ,,
20.	,, 4, ,,	Do. Summit of	2	0·150 ,,	12·975 ,,	14·110 ,,	-1·135 ,,
21.	,, 9, ,,	Do. Second Camp	4	1·180 ,,	15·372 ,,	16·468 ,,	-1·096 ,,
22.	,, 14, ,,	Tambo of Chuquipoquio	4	1·305 ,,	18·575 ,,	19·670 ,,	-1·095 ,,
23.	,, 22, ,,	Ambato	4	1·160 ,,	21·172 ,,	22·094 ,,	-0·922 ,,
24.	,, 29, ,,	Machachi	4	1·150 ,,	20·121 ,,	21·117 ,,	-0·996 ,,
25.	Feb. 8, ,,	Hacienda de la Rosario	4	1·200 ,,	19·810 ,,	20·785 ,,	-0·975 ,,
26.	,, 9, ,,	Illiniza, Camp on S. side	4	1·340 ,,	16·217 ,,	17·239 ,,	-1·022 ,,
27.	,, 16, ,,	Cotopaxi, do.	4	1·310 ,,	16·440 ,,	17·431 ,,	-0·991 ,,
28.	,, 26, ,,	Machachi	4	1·190 ,,	20·075 ,,	21·121 ,,	-1·046 ,,
29.	Mar. 20, ,,	Quito	4	1·175 ,,	20·556 ,,	21·565 ,,	-1·009 ,,
30.	,, 27, ,,	Quebrada of Guallabamba (top)	2	1·200 ,,	20·540 ,,	21·692 ,,	-1·152 ,,
31.	,, ,, ,,	Do. do. (bottom)	2	1·200 ,,	22·800 ,,	23·929 ,,	-1·129 ,,
32.	May 15, ,,	Quito	4	1·080 ,,	20·655 ,,	21·624 ,,	-0·969 ,,
33.	June 8, ,,	Illiniza, Camp on N. side	3	1·410 ,,	16·197 ,,	17·222 ,,	-1·025 ,,
34.	,, 11, ,,	Machachi	4	1·210 ,,	20·142 ,,	....	....

No.	Date.	Place of Observation.	Mercurial Barometer, No. 558 (red. to 32° Faht.)	Aneroid A.	Aneroid E.	Error of A.	Error of E.
1.	Oct. 25, 1879	London	29·768 inch.	29·940 inch.	29·950 inch.	+0·172 inch.	+0·182 inch.
2.	Dec. 2, "	Colon, Isthmus of Panama.	29·854 "	29·810 "	29·800 "	-0·044 "	-0·054 "
3.	" 10, "	Guayaquil, Ecuador	29·767 "	29·750 "	29·750 "	-0·017 "	-0·017 "
4.	" 16, "	Muñapamba, do.	28·611 "	28·525 "	28·500 "	-0·086 "	-0·111 "
5.	" 17, "	Tambo Gobierno, Ecuador	20·777 "	20·640 "	20·400 "	-0·137 "	-0·377 "
6.	" 18, "	Guaranda, do.	21·938 "	21·700 "	21·500 "	-0·238 "	-0·438 "
7.	" 25, "	Do. do.	21·901 "	21·600 "	21·440 "	-0·301 "	-0·461 "
8.	" 27, "	Chimborazo, First Camp	17·872 "	17·600 "	17·220 "	-0·272 "	-0·652 "
9.	" 28, "	Do. Second do.	16·476 "	16·030 "	15·760 "	-0·446 "	-0·716 "
10.	" 30, "	Do. do. do.	16·480 "	15·975 "	15·700 "	-0·505 "	-0·780 "
11.	Jan. 4-5, 1880	Do. Third do.	16·019 "	15·435 "	15·265 "	-0·584 "	-0·754 "
12.	" 4, "	Do. Summit of	14·110 "	...	12·900 "	...	-1·210 "
13.	" 9, "	Do. Second Camp.	16·468 "	15·880 "	15·460 "	-0·588 "	-1·008 "
14.	" 14, "	Tambo of Chuquipoquio	19·683 "	19·125 "	18·775 "	-0·558 "	-0·908 "
15.	" 22, "	Ambato	22·091 "	21·640 "	21·400 "	-0·451 "	-0·691 "
16.	" 29, "	Machachi	21·114 "	20·580 "	20·325 "	-0·534 "	-0·789 "
17.	Feb. 8, "	Hacienda de la Rosario	20·805 "	20·300 "	20·000 "	-0·505 "	-0·805 "
18.	" 9, "	Illiniza, Camp on S. side	17·239 "	16·740 "	16·380 "	-0·499 "	-0·859 "

No.	Date.	Place of Observation.	Mercurial Barometer, No. 558 (red. to 32° Faht.)	Aneroid A.	Aneroid E.	Error of A.	Error of E.
19.	Feb. 16, 1880	Cotopaxi, Camp on	17·431 inch.	16·960 inch.	16·600 inch.	-0·471 inch.	-0·831 inch.
20.	„ 26, „	Machachi	21·142 „	20·550 „	20·290 „	-0·592 „	-0·852 „
21.	Mar. 20, „	Quito	21·577 „	21·000 „	20·800 „	-0·577 „	-0·777 „
22.	May 15, „	Do.	21·631 „	21·070 „	20·890 „	-0·561 „	-0·741 „
23.	„ 20, „	Do.	21·612 „	21·050 „	20·880 „	-0·562 „	-0·732 „
24.	June 8, „	Illiniza, Camp on N. side	17·222 „	16·810 „	...	-0·412 „	...
25.	„ 28, „	Carihuairazo, Camp on	18·545 „	18·030 „	17·690 „	-0·515 „	-0·855 „
26.	„ 29, „	Do. Summit of	16·514 „	16·035 „	15·700 „	-0·479 „	-0·814 „
27.	July 3, „	Chimborazo, Fifth Camp	16·931 „	16·300 „	16·060 „	-0·631 „	-0·871 „
28.	„ „	Do. Summit of	14·044 „	beyond its range	12·990 „	...	-1·054 „
29.	„ 10, „	Commencement of descent towards Pacific Ocean		19·600 „	19·380 „	...	...
30.	„ „	Camp in forest	...	21·250 „	21·200 „	...	...
31.	„ 11, „	Hacienda of Cayandeli	...	24·850 „	24·770 „	...	...
32.	„ „	Camp near Bridge of Chimbo	...	27·930 „	27·910 „	...	...
33.	„ 16, „	Guayaquil	29·911 „	29·550 „	29·590 „	-0·361 „	-0·321 „
34.	„ 27, „	Do.	29·941 „	29·600 „	29·650 „	-0·341 „	-0·291 „
35.	Jan. 9, 1885	London	29·740 „	29·540 „	29·900 „	-0·200 „	+0·160 „

prolonged periods. [The balloon test is only a repetition of the air-pump test. In the former case the instruments are exposed to a natural, and in the latter case to an artificial diminution of pressure ; and if the duration of time is equal in each case the results ought to correspond exactly.]

G. That very material errors may be fallen into by regarding 'a good return' at the level of the sea as a proof of correct working, at low pressures, of aneroids of the present construction.

H. That for the detection of such errors as aneroids (of the present construction) will exhibit when subjected to low pressures for a length of time, aneroids should be subjected artificially to similar pressures for a long period.

PART 2.—EXPERIMENTS IN THE WORKSHOP.

Throughout the second part of this paper, as in the first part, my remarks are principally confined to comparisons of the aneroid against the mercurial barometer. I enter only incidentally into consideration of uses to which these instruments can be put, and do not attempt to explain the mechanical imperfections that cause the errors to which reference is made. The question which I attempt to answer is, Do aneroids, under ordinary conditions, read truly against the mercurial barometer ?

The earliest experiments which were made in the workshop were undertaken with the view of confirming or upsetting the conclusions which were arrived at from observation in the field (§ 18), and they were directed first of all to learn whether it is a fact that *all* aneroids lose upon the mercurial barometer [1] if submitted to diminished pressure for a length of time ; to observe the length of time during which the loss continues to augment, and to ascertain the extent of the loss that occurs.

These earliest experiments were made entirely with aneroids of Mr. J. J. Hicks' manufacture, and they confirmed the observations made in the field. I felt, however, that it was desirable to include instruments by other makers, and I continued to experiment until 70 aneroids had been submitted to examination, and could be reported upon.[2]

Every aneroid which was tested, without exception, lost upon the mercurial barometer when submitted to diminished pressure for a length of time —that is to say, for a day, a week, a month and upwards. It was found that the greater part of such loss as occurred took place during the first week. The experiments were then continued to endeavour to learn what proportion of the loss which occurred during the first week took place during the first *day*.

When this class of experiment had been continued for many months, I turned my attention to the behaviour of aneroids upon their being allowed to return to normal pressure after they had experienced diminished pressure for a length of time. It was found that they always recovered a large part of their loss, and sometimes gained more than they had previously lost. It was found that the recovery might extend over several weeks, and that the greater part of the recovery or gain usually occurred in the first week. As the amount which was recovered very seldom exactly equalled the amount of the previous loss, in the great majority of cases there was a marked change in the index-errors, and, in some instances, a *large* alteration.

[1] See § 21 for explanation of the expression "loss upon the mercurial barometer."
[2] A number of others were tested. Some behaved well, but they are not included here, as they did not bear makers' names.

I then endeavoured to learn whether aneroids which possessed large index-errors could nevertheless be usefully employed for the measurement of variations of pressure, and found that in the majority of cases they might be so employed.   Other experiments having a practical bearing suggested themselves from time to time.

§ 19. The method by which the aneroids were tested in these experiments was that which is habitually employed during 'verification' of these instruments.   The aneroids were placed under the receiver of an air-pump, to which there was an attached mercurial barometer.   Upon air being withdrawn, there was simultaneous reduction in pressure both for the aneroids and for the mercurial.   The indications of the latter were frequently checked by reference to a standard mercurial barometer, hanging alongside.

§ 20. Although my method of testing was the same as that employed during 'verification,' there was one essential point of difference between the verification tests and my own, namely in the length of time during which the aneroids were kept at reduced pressures.   I have good authority for saying that even when an aneroid is verified at Kew Observatory inch by inch, down to as low as 15 inches, it is unusual to occupy more than an hour in the operation,—about one-half of which amount of time will be consumed whilst pressure is being reduced, and the rest while pressure is being restored.

This, also, is about the length of time ordinarily occupied during the manufacture of aneroids upon the process termed 'pointing,' *i.e.* laying off the scale on the dial of an aneroid (by comparison with an attached mercurial barometer) prior to graduation.   Therefore, what verification amounts to is this.   It is a repetition of the temporary reduction of pressure to which aneroids have been subjected in the course of manufacture ; and one learns from verification whether the 'pointing' and the subsequent graduation have been accurately performed.   One does not learn from it the errors that will be manifested by aneroids which may be subjected to a reduction of pressure *for a greater length of time.*   In order that my tests in the workshop might be the equivalent of the tests to which aneroids are put in the field I kept the instruments at various pressures between 26 and 14 inches for periods of days, weeks, and even months at a time.

§ 21. The first six aneroids that were experimented upon were taken from the stock of Mr. Hicks, and were marked temporarily 1-6. These were placed under the receiver, and had pressure reduced to 22·5 inches, and were kept continuously at that pressure for six weeks.   The annexed table scarcely requires explanation.   In Column 1 the errors of the aneroids are given which were exhibited upon their being reduced to 22·5 inches.   In Columns 2-7 the errors are given which they showed at the end of each week, during six successive weeks ; and in Column 8 the total amount of the loss is stated that occurred during the six weeks.   In the case of aneroid No. 1, a + error of 0·159 was converted into a − error of 0·335.   The actual loss amounted therefore to 0·494 of an inch.   In the case of No. 5, the − error of 0·016 was increased to − 0·485.   The actual loss was therefore 0·469 of an inch.   The amounts so lost I term "loss upon the mercurial barometer."

EXPERIMENT IN WHICH SIX (WATCH-SIZE) ANEROIDS WERE KEPT AT A PRESSURE OF 22·5 INCHES FOR SIX WEEKS.

Aneroid.	1. Errors of aneroids at 22·5 inches at the start of the experiment.	2. Errors of aneroids at the end of the first week.	3. Errors of aneroids at the end of the second week.	4. Errors of aneroids at the end of the third week.	5. Errors of aneroids at the end of the fourth week.	6. Errors of aneroids at the end of the fifth week.	7. Errors of aneroids at the end of the sixth week.	8. Loss of aneroids upon the Merc. Bar. in six weeks.
	inch.	inch.	inch.	inch.	inch.	inch.	inch.	inch.
No. 1. (Hicks)	+ 0·159	− 0·267	− 0·291	− 0·321	− 0·347	− 0·342	− 0·335	0·494
,, 2. ,,	+ 0·059	− 0·267	− 0·301	− 0·331	− 0·352	− 0·347	− 0·345	0·404
,, 3. ,,	− 0·011	− 0·167	− 0·191	− 0·211	− 0·202	− 0·207	− 0·220	0·209
,, 4. ,,	+ 0·059	− 0·197	− 0·211	− 0·251	− 0·272	− 0·267	− 0·265	0·324
,, 5. ,,	− 0·016	− 0·387	− 0·431	− 0·441	− 0·462	− 0·472	− 0·485	0·469
,, 6. ,,	+ 0·039	− 0·217	− 0·266	− 0·301	− 0·297	− 0·297	− 0·300	0·339
Mean errors of aneroids on Mercurial Barometer (reduced to 32° Faht.)	+ 0·048	− 0·250	− 0·282	− 0·309	− 0·322	− 0·322	− 0·325	0·373
Greatest differences of the aneroids	0·175	0·220	0·240	0·230	0·260	0·265	0·265	....

Note. In calculations for altitude, the value of 0·373 of an inch, at a pressure of 22·5 inches, is about 450 feet.

D

Several points came out very clearly during this experiment. The first and principal one was that the whole six instruments lost considerably upon the mercurial barometer. Another, and it seemed to me an important one, was that the greater part of the loss occurred in each instance *during the first week.* In every succeeding experiment these facts were confirmed. Another point was the large difference in the loss in the different instruments—the least being 0·209 and the greatest 0·494 of an inch. The gradual stoppage in increase of the loss is best seen by examining the line in which the *mean* errors of the whole are given. At the end of the fifth week there was no perceptible increase in the mean error, but at the end of the sixth week a slight increase appeared. This was due to Nos. 3 and 5, which probably would have continued to lose some small amounts for several weeks longer.

§ 22. The emphatic manner in which this first experiment in the workshop confirmed the experiences in the field keenly interested those with whom I was associated, and immediately upon its termination we started a fresh series of aneroids (marked temporarily 10–12) at a pressure of 17 inches, and kept them at that pressure during five weeks, with the following results.

EXPERIMENT IN WHICH THREE ANEROIDS (NO. 10, WATCH-SIZE ; NOS. 11, 12, THREE INCHES DIAMETER) WERE KEPT AT A PRESSURE OF 17 INCHES FOR FIVE WEEKS.

Aneroid.	1. Errors of aneroids at 17 inches at the start of the experiment.	2. Errors of aneroids at the end of the first week.	3. Errors of aneroids at the end of the fifth week.	4. Loss of aneroids upon the Merc. Bar. in five weeks.
	inch.	inch.	inch.	inch.
No. 10 (Hicks) .	− 0·023	− 0·731	− 0·944	0·921
„ 11 „	− 0·203	− 0·721	− 0·909	0·706
„ 12 „	− 0·043	− 0·531	− 0·709	0·666
Mean errors of aneroids on Mercurial Barometer (reduced to 32° Faht.)	− 0·090	− 0·661	− 0·854	0·764

Note. In calculations for altitude, the value of 0·764 of an inch, at a pressure of 17 inches, is about 1220 feet.

I do not feel it necessary to give this second experiment in as full detail as the previous one. Each aneroid lost considerably upon the mercurial barometer ; in each case the greater part of the loss occurred in the first week ; the loss was different in each instrument ; and the loss seemed to cease to augment about the fourth or fifth week. No. 10 lost nearly an inch (0·921), and the mean loss of the three instruments amounted to 0·764 of an inch.

§ 23. When this experiment was concluded I started a third series of aneroids (Nos. 7, 8, 9), three inches diameter each, at a pressure of 16 inches, and kept them at that pressure for two months. This series lost less at 16 inches during six weeks than Nos. 10, 11, and 12 had lost at 17 inches during five weeks. But the loss, as before, was considerable in each instrument ; the greater part of the loss occurred during the first week ; and the loss was comparatively trifling after the fourth week.

EXPERIMENT IN WHICH THREE ANEROIDS (EACH THREE INCHES DIAMETER) WERE KEPT AT A PRESSURE OF 16 INCHES DURING EIGHT WEEKS.

Aneroid.	1. Errors of aneroids at 16 inches at start of the experiment.	2. Errors of aneroids at the end of the first week.	3. Errors of aneroids at the end of the second week.	4. Errors of aneroids at the end of the third week.
	inch.	inch.	inch.	inch.
No. 7 (Hicks) .	+ 0·104	− 0·118	− 0·180	− 0·206
„ 8 „ . .	+ 0·069	− 0·223	− 0·305	− 0·306
„ 9 „ . .	+ 0·014	− 0·368	− 0·405	− 0·451
Mean errors of aneroids on Mercurial Barometer (reduced to 32° Faht.)	+ 0·062	− 0·236	− 0·297	− 0·321

5. Errors of aneroids at the end of the fourth week.	6. Errors of aneroids at the end of the fifth week.	7. Errors of aneroids at the end of the sixth week.	8. Errors of aneroids at the end of the seventh week.	9. Errors of aneroids at the end of the eighth week.	10. Loss of aneroids upon the Merc. Bar. in eight weeks.
inch.	inch.	inch.	inch.	inch.	inch.
− 0·252	− 0·276	− 0·285	− 0·306	− 0·271	0·375
− 0·322	− 0·326	− 0·355	− 0·331	− 0·336	0·405
− 0·472	− 0·476	− 0·495	− 0·506	− 0·531	0·545
− 0·349	− 0·359	− 0·378	− 0·381	− 0·379	0·442

Note. In calculations for altitude, the value of 0·442 of an inch, at a pressure of 16 inches, is about 750 feet.

§ 24. After this experiment, I proceeded to test four aneroids which had been specially made, and embraced some peculiarities in construction. In this (the fourth) series each instrument lost considerably upon the mercurial barometer; the greater part of the loss occurred during the first week; and the mean error of the whole four ceased to augment after the fourth week.

EXPERIMENT IN WHICH FOUR ANEROIDS (EACH FOUR AND A HALF INCHES DIAMETER) WERE KEPT AT A PRESSURE OF 19 INCHES DURING FIVE WEEKS.

Aneroids 13, 14, 15, 16 (Hicks).	Mean error of the aneroids at 19 inches at start of the experiment.	Mean error of the aneroids at the end of the first week.	Mean error of the aneroids at the end of the fifth week.	Mean loss upon Mercurial Barometer in five weeks.
	inch.	inch.	inch.	inch.
Mean errors of the four aneroids on the Mercurial Barometer (reduced to 32° Faht.)	+ 0·013	− 0·246	− 0·348	0·361

Note. In calculations for altitude, the value of 0·361 of an inch, at a pressure of 19 inches, is about 520 feet.

§ 25. These four experiments (as well as subsequent ones) showed clearly that the greater part of the loss which occurred took place during the first week. From the following table it will be seen that in every series the mean loss in a week exceeded two-thirds of the total mean loss upon the mercurial barometer.

Number of Aneroids employed.	Pressures at which they were kept.	Length of time during which they were kept at these pressures.	Mean loss of each series upon the Merc. Bar. in one week.	Total mean loss of each series upon the Mercurial Barometer.
	inches.		inch.	inch.
6	22·5	6 weeks	0·298	0·373
3	17	5 „	0·571	0·764
3	16	8 „	0·298	0·442
4	19	5 „	0·259	0·361

§ 26. As it appeared from these experiments that the greater (and more important) part of the loss which occurred took place in the first *week*, I continued by testing aneroids, for one week each, at every inch of the barometer between 14 and 26 inches, and endeavoured to procure instruments by a diversity of makers and of various diameters. This series of experiments necessarily extended over several years, as it was seldom possible to test more than a few instruments at a time, owing to difficulty in obtaining at any one time a number of instruments with similar ranges. It included aneroids by Hicks, Casella, Adie, Elliott, Negretti and Zambra, Cooke and Sons, Secretan, and Hilger.

No.	Maker's name and mark.	Property of	Diameter of instrument.	Pressure at which it was kept during one week.	Approximate equivalent altitude.	Loss upon the attached Mercurial Bar. in one week.	Approximately equal to a + error in height of
			inches.	inches.	feet.	inches.	feet.
1.	J. J. Hicks (Watkin Patent, 9)	The Maker	4½	26	3900	0·087	90
2.	do. ( ) do. 141)	do.	3	,,	,,	0·094	98
3.	do. ( ) do. 165)	do.	4½	,,	,,	0·186	194
4.	L. Casella . (2923)	do.	4½	,,	,,	0·132	137
5.	Negretti and Zambra . (115)	Meteorological Office	4½	25	4970	0·070	76
6.	Elliott (419)	do.	4½	,,	,,	0·115	126
7.	Adie (38)	do.	4½	,,	,,	0·125	137
8.	J. J. Hicks (Watkin Patent, 17)	The Maker	4½	,,	,,	0·144	156
9.	do. (33)	do.	2	,,	,,	0·175	192
10.	do. (32)	do.	2	,,	,,	0·185	201
11.	L. Casella (4493)	do.	4	24·5	5520	0·185	205
12.	do. (1842)	do.	4½	,,	,,	0·200	222
13.	do. (5682)	do.	3	,,	,,	0·245	272
14.	Secretan —	F. F. Tuckett	4½	24	6080	0·067	76
15.	J. J. Hicks (31)	The Maker	2	,,	,,	0·070	80
16.	Adie (33)	Meteorological Office	4½	,,	,,	0·133	151
17.	J. J. Hicks (28)	The Maker	2	,,	,,	0·143	162
18.	T. Cooke and Sons (170)	F. F. Tuckett	4	,,	,,	0·153	174

No.	Maker's name and mark.	Property of	Diameter of instrument.	Pressure at which it was kept during one week.	Approximate equivalent altitude.	Loss upon the attached Mercurial Bar. in one week.	Approximately equal to a + error in height of
			inches.	inches.	feet.	inches.	feet.
19.	J. J. Hicks (30)	The Maker	2	24	6080	0·200	227
20.	do. (27)	do.	2	"	"	0·203	231
21.	Negretti and Zambra —	Meteorological Office	4½	"	"	0·223	253
22.	J. J. Hicks (35)	The Maker	2	"	"	0·265	301
23.	Negretti and Zambra —	Meteorological Office	2¾	"	"	0·266	302
24.	J. J. Hicks (Watkin Patent, 223)	The Maker	3	"	"	0·273	309
25.	do. (29)	do.	2	"	"	0·295	335
26.	do. (26)	do.	2	"	"	0·338	384
27.	do. (20)	do.	4½	23	7240	0·252	300
28.	T. Cooke and Sons (160)	R. Etheridge	2¾	"	"	0·264	314
29.	J. J. Hicks (3)	The Maker	2	22·5	7840	0·156	186
30.	do. (4)	do.	2	"	"	0·256	305
31.	do. (6)	do.	2	"	"	0·256	305
32.	do. (2)	do.	2	"	"	0·326	390
33.	do. (5)	do.	2	"	"	0·371	442
34.	do. (1)	do.	2	"	"	0·426	507
35.	L. Casella (5893)	do.	2	22	8450	0·353	441
36.	Hilger —	Ed. Whymper	2	"	"	0·458	572

No.	Maker's name and mark.	Property of	Diameter of instrument.	Pressure at which it was kept during one week.	Approximate equivalent altitude.	Loss upon the attached Mercurial Bar. in one week.	Approximately equal to a + error in height of
			inches.	inches.	feet.	inches.	feet.
37.	L. Casella . (5793)	The Maker .	2	22	8,450	0·473	591
38.	J. J. Hicks . (F)	Ed. Whymper	2	21·5	9,080	0·590	737
39.	L. Casella . (580)	do.	2	21	9,720	0·461	591
40.	do. . (1021)	do.	2¾	,,	,,	0·533	683
41.	J. J. Hicks . (42)	The Maker	2	20	11,050	0·287	387
42.	do. . (40)	do.	2	,,	,,	0·337	456
43.	do. . (38)	do.	2	,,	,,	0·362	489
44.	L. Casella . (5892)	do.	2	,,	,,	0·652	880
45.	J. J. Hicks . (37)	do.	2	,,	,,	0·677	915
46.	L. Casella . (5792)	do.	2	,,	,,	0·767	1037
47.	J. J. Hicks . (39)	do.	2	,,	,,	0·842	1137
48.	do. . (14)	do.	4½	19	12,450	0·120	172
49.	do. . (16)	do.	4½	,,	,,	0·290	418
50.	do. . (15)	do.	4½	,,	,,	0·300	430
51.	do. . (13)	do.	4½	,,	,,	0·325	468
52.	L. Casella . (5606)	do.	2	18	13,920	0·463	707
53.	do. . (4885)	C. Dent .	2	,,	,,	0·474	734
54.	do. . (5798)	The Maker	3	,,	,,	0·653	997

No.	Maker's name and mark.	Property of	Diameter of instrument.	Pressure at which it was kept during one week.	Approximate equivalent altitude.	Loss upon the attached Mercurial Bar. in one week.	Approximately equal to a + error in height of
			inches.	inches.	feet.	inches.	feet.
55.	Cary . . . (868)	Royal Geog. Soc.	2¾	17	15,480	0·322	515
56.	do. . . (732)	do.	2¾	,,	,,	0·362	579
57.	J. J. Hicks . (12)	The Maker	3	,,	,,	0·488	780
58.	do. (11)	do.	3	,,	,,	0·518	829
59.	Cary . . (829)	Royal Geog. Soc.	2¾	,,	,,	0·537	859
60.	J. J. Hicks . (10)	The Maker	2	,,	,,	0·708	1133
61.	L. Casella . (5606)	do.	2	,,	,,	0·727	1163
*62.	Cary . (841)	Royal Geog. Soc.	2¾	,,	,,	1·267	2027
63.	J. J. Hicks . (7)	The Maker	3	16	17,130	0·222	378
64.	do. . . (8)	do.	3	,,	,,	0·292	500
65.	do. . . (9)	do.	3	,,	,,	0·382	653
†66.	J. J. Hicks (Watkin Patent, 90)	H. Woolley	3	,,	,,	0·405	692
†67.	Cary . . . (899)	do.	2¾	,,	,,	0·625	1068
68.	J. J. Hicks . (22)	The Maker	3	15	18,890	0·701	1274
69.	do. . . (23)	do.	3	,,	,,	0·746	1356
70.	do. . . (E)	Ed. Whymper	2	14	20,770	0·803	1559
71.	do. . . (F)	do.	2	,,	,,	1·033	2005

* The aneroid employed by Mr. T. Thomson on his journey in Morocco, in 1888.   † Used during journeys in the Caucasus.

These tables speak for themselves. I draw attention, however, to the last column of each, in which there is shown approximately the errors in determinations of altitude which would have been probable if these particular aneroids had been employed in the field, at the mentioned pressures, for *one week* ; and I point out especially that these errors, serious as they are, are (taking them as a whole) probably only about *two-thirds* of the maximum errors which these particular aneroids would have developed, if they had been kept continuously at the mentioned pressures for one month and upwards.

§ 27. It appears to me to follow, and to be indubitable, that a great part of the altitudes throughout the world which depend upon observations of aneroid barometers made while ascending must be *too high*, and that a general lowering of them will be found necessary ; and that it is probable the reduction in height will have to be at a greater rate *per cent* in the case of the loftier positions than in the case of the inferior ones, and with those which are in the interiors of continents than with those in the neighbourhood of the sea-level.

§ 28. Concurrently with these observations to attempt to learn the extent of the loss of aneroids upon the mercurial barometer on being submitted to diminished pressures for a *week*, I observed the loss which occurred in the first *day* of the first week. This is variable,—sometimes being about *one-third* of the week's loss,[1] and sometimes *more than three-fourths* of it. This is illustrated by the examples which are given on p. 26.

§ 29. The experiments which have been already quoted indicate that the loss in aneroids upon the mercurial barometer augments at a constantly diminishing rate, provided they are kept continuously at the same pressure. The loss is greater in the first *week* than it is in the second, or in any succeeding week. It is rare, however, to observe in any seven successive *days* a perfectly regular (or symmetrical) increase in the errors. Minor imperfections of construction prevent perfectly harmonious readings, and render it impossible to say more than that the loss during the first *day* of the first week (so far as my observation extends) is almost always greater than during any subsequent one in the first week. A few examples are given upon p. 27 in illustration.

§ 30. The loss in aneroids upon the mercurial barometer can be observed in almost all instruments during the first *hour* they are subjected to diminished pressure, if the diminution in pressure amounts to several inches ; and in aneroids giving the inch of the mercurial barometer that length upon their scales, and more particularly upon aneroids with expanded scales (Hicks' Watkin Patent) the loss may even be traced in successive hours.[2] The loss in the first hour of the first day, so far as my observation extends, is always greater than in any subsequent one.

---

[1] I have observed some exceptional cases (not quoted in the table) in which the day's loss has been *less than a fifth* of the week's loss.

[2] When standing over these instruments, whilst they have been kept at so moderate a diminution in pressure as 26 inches, I have seen the index move backwards whilst the mercurial barometer remained immovable.

E

TABLE SHOWING THE AMOUNTS LOST UPON THE MERCURIAL BAROMETER
IN ONE *DAY* AND IN ONE *WEEK*.

Aneroid.		Diameter of instrument.	Pressure at which it was kept.	Loss upon the Mercurial Bar. in one day.	Total loss upon the Merc. Bar. in one week.
		inch.	inch.	inch.	inch.
Hicks . . .	(E)	2	14	0·510	0·803
do. . . .	(F)	2	,,	0·720	1·033
do. . . .	(22)	3	15	0·546	0·701
do. . . .	(23)	3	,,	0·481	0·746
Casella . . .	(5606)	2	17	0·521	0·727
do. . . .	(5606)	2	18	0·378	0·463
do. . . .	(5798)	3	,,	0·513	0·653
do. . . .	(5792)	2	20	0·660	0·767
do. . . .	(5892)	2	,,	0·545	0·652
Hicks . . .	(37)	2	,,	0·545	0·677
do. . . .	(38)	2	,,	0·280	0·362
do. . . .	(39)	2	,,	0·685	0·842
do. . . .	(40)	2	,,	0·255	0·337
do. . . .	(42)	2	,,	0·225	0·287
Casella . . .	(580)	2	21	0·325	0·461
do. . . .	(1021)	$2\frac{3}{4}$	,,	0·385	0·533
do. . . .	(5793)	2	22	0·342	0·473
do. . . .	(5893)	2	,,	0·227	0·353
Hilger . . .	—	2	,,	0·292	0·458
Hicks . . .	(20)	$4\frac{1}{2}$	23	0·154	0·252
Cooke . . .	(160)	$2\frac{3}{4}$	,,	0·161	0·264
Hicks (Watkin Patent, 232)		3	24	0·111	0·273
Casella . . .	(4493)	4	24·5	0·146	0·185
do. . . .	(5682)	3	,,	0·181	0·245
do. . . .	(1842)	$4\frac{1}{2}$	,,	0·121	0·200
Hicks (Watkin Patent, 17)		$4\frac{1}{2}$	25	0·052	0·144
do. (do. 9)		$4\frac{1}{2}$	26	0·069	0·087
do. (do. 141)		3	,,	0·084	0·094
Casella . . .	(2923)	$4\frac{1}{2}$	,,	0·099	0·132

TABLE SHOWING THE INCREASE IN THE LOSS UPON THE MERCURIAL BAROMETER FROM THE END OF THE FIRST DAY TO THE END OF THE SEVENTH DAY.

Aneroid.	Diameter of Instrument.	Pressure at which it was kept.	Error on the Merc. Bar. (red. to 32°) upon being reduced to the pressure at which it was kept.	Error on the Merc. Bar. at the end of the first day.	Error on the Merc. Bar. at the end of the second day.	Error on the Merc. Bar. at the end of the fourth day.	Error on the Merc. Bar. at the end of the seventh day.
	inch.	inch.	inch.	inch.	inch.	inch.	inch.
Hicks . . . (E)	2	14	− 0·208	− 0·718	− 0·769	− 0·860	− 1·011
do. . . . (F)	2	,, 15	− 0·158	− 0·878	− 0·964	− 1·060	− 1·191
do. . . . (22)	3	,,	+ 0·047	− 0·499	− 0·560	− 0·599	− 0·654
do. . . . (23)	3	,, 20	+ 0·247	− 0·234	− 0·340	− 0·399	− 0·499
do. . . . (37)	2	,,	− 0·164	− 0·709	not observed	− 0·806	− 0·841
do. . . . (39)	2	,,	− 0·374	− 1·059	,,	− 1·181	− 1·216
Casella . (5792)	2	,,	+ 0·456	− 0·204	,,	− 0·271	− 0·311
do. . . (5892)	2	,,	+ 0·066	− 0·479	,,	− 0·556	− 0·586
do. . . (580)	2	,, 21	+ 0·384	+ 0·059	+ 0·006	− 0·056	− 0·077
do. . . (1021)	2¾	,,	− 1·046	− 1·431	− 1·474	− 1·536	− 1·579
do. . . (5793)	2	,, 22	+ 0·094	− 0·248	− 0·291	− 0·334	− 0·379
do. . . (5893)	2	,,	+ 0·079	− 0·148	− 0·191	− 0·234	− 0·274
Hilger . . —	2	,,	+ 0·339	+ 0·047	− 0·006	− 0·074	− 0·119
Cooke . . (160)	2¾	,, 23	+ 0·353	+ 0·192	+ 0·173	+ 0·125	+ 0·089
Hicks . . (20)	4½	,,	+ 0·056	− 0·098	− 0·117	− 0·160	− 0·196
Hicks (Watkin Patent, 232)	3	,, 24	+ 0·118	+ 0·007	− 0·058	− 0·095	− 0·155
Casella . (4493)	4	24·5	+ 0·049	− 0·097	− 0·109	− 0·127	− 0·136
do. . . (5682)	3	,,	+ 0·014	− 0·167	− 0·199	− 0·202	− 0·231
do. . . (1842)	4½	,,	+ 0·084	− 0·037	− 0·069	− 0·092	− 0·116
do. . . (2923)	4½	,, 26	+ 0·155	+ 0·056	+ 0·042	+ 0·033	+ 0·023
Hicks (Watkin Patent, 9)	4½	,,	+ 0·065	− 0·004	− 0·008	− 0·007*	− 0·022

* Discordant.

TABLE SHOWING THE AMOUNTS LOST UPON THE MERCURIAL BAROMETER IN THE FIRST HOUR AND IN THE FIRST DAY.

Aneroid.	Diameter of instrument.	Pressure at which it was kept.	Error on the Merc. Bar. (red. to 32°) upon being reduced to the pressure at which it was kept.	Error on the Merc. Bar. at the end of the first hour.	Error on the Merc. Bar. at the end of the first day.
	inch.	inch.	inch.	inch.	inch.
Hicks . . (E)	2	14	− 0·208	− 0·531	− 0·718
do. . . (F)	2	,,	− 0·158	− 0·596	− 0·878
do. . . (22)	3	15	+ 0·047	− 0·274	− 0·499
do. . . (23)	3	,,	+ 0·247	− 0·024	− 0·234
do. . . (11)	3	17	− 0·604	− 0·852	− 1·054
do. . . (12)	3	,,	− 0·234	− 0·437	− 0·604
do. . . (22)	3	18	+ 0·001	− 0·254	− 0·382
do. . . (23)	3	,,	+ 0·196	− 0·004	− 0·132
do. . . (37)	2	20	− 0·164	− 0·526	− 0·709
do. . . (38)	2	,,	− 0·024	− 0·126	− 0·304
do. . . (39)	2	,,	− 0·374	− 0·676	− 1·059
do. . . (40)	2	,,	+ 0·676	+ 0·554	+ 0·421
do. . . (42)	2	,,	− 0·024	− 0·086	− 0·249
Casella . . (5792)	2	,,	+ 0·456	+ 0·124	− 0·204
do. . . (5892)	2	,,	+ 0·066	− 0·201	− 0·479
do. . . (580)	2	21	+ 0·296	+ 0·184	− 0·002
do. . . (1021)	2¾	,,	− 1·174	− 1·331	− 1·512
do. . . (5793)	2	22	+ 0·094	− 0·082	− 0·291
do. . . (5893)	2	,,	+ 0·079	− 0·057	− 0·191
Hilger . . —	2	,,	+ 0·339	+ 0·188	− 0·006
Hicks . . (20)	4½	23	+ 0·056	− 0·052	− 0·098
Cooke . . (160)	2¾	,,	+ 0·353	+ 0·263	+ 0·192
Hicks (Watkin P. 232)	3	24	+ 0·118	+ 0·069	+ 0·007
Hicks . . (32)	2	25	+ 0·274	+ 0·213	+ 0·118
do. . . (33)	2	,,	− 0·121	− 0·172	− 0·207
Casella . . (2923)	4½	26	+ 0·155	+ 0·094	+ 0·056
Hicks (Watkin P. 9)	4½	,,	+ 0·065	+ 0·029	− 0·004
do. (do. 141)	3	,,	+ 0·065	+ 0·014	− 0·019
do. (do. 165)	4½	,,	+ 0·142	+ 0·087	+ 0·024

§ 31. From the foregoing experiments it is clearly apparent that the *extent* of the loss which will occur in any aneroid upon the mercurial barometer, upon being submitted to diminished pressure, depends, 1. upon the duration of time it may be submitted to diminished pressure, and 2. upon the extent of the diminution in pressure ; and it follows that the errors which will be manifested by any particular aneroid will be greatest when it is submitted to very low pressures for long periods.[1] Consideration of some practical conclusions which may be drawn from these facts is deferred to the third portion of this paper, and I now proceed to give some examples of the behaviour of aneroids upon their return to normal pressure, after having experienced diminished pressure for a length of time.

§ 32. The first series which was experimented upon (Nos. 1–6) was watched by me for twenty-one weeks after being allowed to return to the natural pressure of the time ; and it was found that the large mean error which existed at the termination of the experiment, namely − 0·325 of an inch (see table on p. 17), gradually lessened, and that on the expiration of the twelfth week the six aneroids had 'recovered' all they had previously lost. The mean error at the commencement of the experiment was + 0·048, and at the end of the twelfth week (after return to natural pressure) it stood at + 0·057 of an inch.[2] In nine succeeding weeks I was unable to trace any material change in the mean index-error.

The three aneroids Nos. 10, 11, 12, which were employed in the second experiment (see p. 18), were watched by me for sixteen weeks after return to normal pressure. Their very large mean error of − 0·854 was gradually reduced,—at the end of the first week to − 0·223, at the end of the second week to − 0·151, and so on, until at the end of the fifth week it stood at − 0·110 of an inch. After this no farther recovery could be traced.

§ 33. In these two experiments, and in all such succeeding ones as extended over a number of weeks, it was clear that the greater part of the recovery occurred within the first week after return to normal pressure, and I accordingly endeavoured to learn within what space of time the most important part of the loss was recovered. With this view, I watched the recovery of a number of aneroids (of various diameters, by several makers, which had been kept at different pressures) upon each day during a week, and observed the errors of the aneroids upon the mercurial barometer on each successive day. See table upon p. 30.

§ 34. From these observations, I found that the greater part of the recovery which occurred during the first week after return to normal pressure took place in the first day of the week ; and that the amount recovered in the first *hour* of the first day was always considerable, and was almost always greater than the amount still further recovered in any subsequent hour. See table upon p. 31.

---

[1] The loss is not strictly proportional throughout the whole length of the scale of any aneroid, and at the last (or lowest) inch of the graduation it is sometimes much greater than at any other part of the scale. *It is not advisable to attempt to work an aneroid to the full extent of its range.*

[2] The most important part of the recovery occurred much earlier.

TABLE SHOWING THE AMOUNTS RECOVERED IN ONE *DAY* AND IN ONE *WEEK.*

Aneroid.		Diameter of Instrument.	Pressure at which it was kept for one week.	**1.** Error on the Merc. Bar. at natural pressure before reduction of pressure.	**2.** Error on the Merc. Bar. on return to normal pressure.	**3.** Error on the Merc. Bar. one day later.	**4.** Error on the Merc. Bar. one week later.	
		inch.	inch.	inch.	inch.	inch.	inch.	
Hicks	(E)	2	14	+ 0·248	− 0·238	+ 0·143	+ 0·213	
do.	(F)	2	„	+ 0·098	− 0·598	+ 0·013	+ 0·098	
do.	(22)	3	15	− 0·036	− 0·360	− 0·050	+ 0·006	*
do.	(23)	3	„	− 0·206	− 0·560	− 0·265	− 0·209	
Cary	(732)	2¾	17	+ 0·142	− 0·379	− 0·213	− 0·198	
do.	(829)	2¾	„	+ 0·182	− 0·539	− 0·288	− 0·223	
do.	(841)	2¾	„	+ 0·442	− 1·029	− 0·163	+ 0·052	
Casella	(5606)	2	18	+ 0·158	− 0·220	+ 0·062	+ 0·090	
do.	(4885)	2	„	+ 0·113	− 0·087	+ 0·130	+ 0·174	*
do.	(5798)	3	„	+ 0·058	− 0·430	− 0·113	− 0·075	
Cooke	(170)	4	„	− 0·237	− 0·537	− 0·295	− 0·201	*
Secretan	—	4½	„	+ 0·029	− 0·028	+ 0·057	+ 0·068	*
Hicks	(37)	2	20	+ 0·080	− 0·547	− 0·092	+ 0·028	
do.	(38)	2	„	− 0·065	− 0·407	− 0·187	− 0·092	
do.	(39)	2	„	− 0·140	− 0·977	− 0·457	− 0·317	
do.	(40)	2	„	+ 0·415	+ 0·263	+ 0·558	+ 0·708	*
do.	(42)	2	„	− 0·105	− 0·272	− 0·162	− 0·072	*
Casella	(5792)	2	„	+ 0·245	− 0·237	+ 0·068	+ 0·158	
do.	(5892)	2	„	+ 0·110	− 0·222	− 0·052	+ 0·043	
do.	(580)	2	21	+ 0·351	− 0·098	+ 0·199	+ 0·298	
do.	(1021)	2¾	„	− 1·249	− 1·848	− 1·526	− 1·332	
Hilger	—	2	22	+ 0·180	− 0·281	+ 0·013	+ 0·054	
Casella	(5793)	2	„	+ 0·010	− 0·356	− 0·087	− 0·021	
do.	(5893)	2	„	+ 0·180	− 0·056	+ 0·083	+ 0·129	
Cooke	(160)	2¾	23	− 0·036	− 0·191	− 0·049	− 0·025	*
Hicks	(20)	4½	„	+ 0·014	− 0·166	− 0·069	− 0·013	
Hicks (Watkin P. 232)		3	24	− 0·022	− 0·169	− 0·105	− 0·074	
Hicks	(29)	2	25·2	− 0·209	− 0·507	− 0·414	− 0·310	
do.	(31)	2	„	+ 0·166	+ 0·023	+ 0·136	+ 0·160	
do.	(32)	2	„	+ 0·161	− 0·017	+ 0·056	+ 0·200	*
do.	(33)	2	„	− 0·179	− 0·277	− 0·209	− 0·200	
do.	(35)	2	„	+ 0·266	+ 0·203	+ 0·306	+ 0·345	*
Hicks (Watkin P. 9)		4½	26	+ 0·013	− 0·045	+ 0·048	+ 0·059	*
do. (do. 141)		3	„	− 0·037	− 0·155	− 0·050	− 0·020	*
Casella	(2923)	4½	„	+ 0·093	+ 0·005	+ 0·095	+ 0·103	*
Hicks (Watkin P. 165)		4½	„	+ 0·039	− 0·163	− 0·061	+ 0·005	

TABLE SHOWING THE AMOUNTS RECOVERED IN ONE *HOUR* AND IN ONE *DAY*.

Aneroid.	Diameter of instrument.	Pressure at which it was kept for one week.	Error on the Merc. Bar. at natural pressure before reduction of pressure.	Error on the Merc. Bar. on return to normal pressure.	Error on the Merc. Bar. one hour later.	Error on the Merc. Bar. one day later.
	inch.	inch.	inch.	inch.	inch.	inch.
Hicks . . (E)	2	14	+ 0·248	− 0·238	+ 0·012	+ 0·143
do. . . (F)	2	,,	+ 0·098	− 0·598	− 0·278	+ 0·013
do. . . (22)	3	15	− 0·036	− 0·360	− 0·250	− 0·050
do. . . (23)	3	,,	− 0·236	− 0·560	− 0·470	− 0·265
Casella . . (5606)	2	18	+ 0·158	− 0·220	− 0·043	+ 0·062
do. . . (5798)	3	,,	+ 0·058	− 0·430	− 0·283	− 0·113
do. . . (5792)	2	20	+ 0·245	− 0·237	− 0·059	+ 0·068
do. . . (5892)	2	,,	+ 0·110	− 0·222	− 0·144	− 0·052
Hicks . . (37)	2	,,	+ 0·080	− 0·547	− 0·234	− 0·092
do. . . (38)	2	,,	− 0·065	− 0·407	− 0·274	− 0·187
do. . . (39)	2	,,	− 0·140	− 0·977	− 0·664	− 0·457
do. . . (40)	2	,,	+ 0·415	+ 0·263	+ 0·291	+ 0·558
do. . . (42)	2	,,	− 0·105	− 0·272	− 0·214	− 0·162
Hilger . . —	2	22	+ 0·180	− 0·281	− 0·084	+ 0·013
Casella . . (5793)	2	,,	+ 0·010	− 0·356	− 0·214	− 0·087
do. . . (5893)	2	,,	+ 0·180	− 0·056	+ 0·016	+ 0·083
Cooke . . (160)	2¾	23	− 0·036	− 0·191	− 0·108	− 0·049
Hicks . . (20)	4½	,,	+ 0·014	− 0·166	− 0·108	− 0·069
Hicks(Watkin P. 232)	3	24	− 0·022	− 0·169	− 0·112	− 0·105
do. (do. 165)	4½	26	+ 0·039	− 0·163	− 0·102	− 0·061

§ 35. From examination of the whole of the observations upon the recovery I find :—

That the greater the length of time during which aneroids may be submitted to a reduction of pressure the greater is the length of time over which their recovery extends.

That, in instances where aneroids have been kept continuously at diminished pressures for a month and upwards, the greater part of the recovery that occurs takes place in the first *week* after return to normal pressure.

That, in instances where aneroids have been kept continuously at diminished pressures for a week, the greater part of the recovery which occurs in the first week after return to normal pressure usually takes place within the first *day*.

That the amount recovered in the first *hour* is almost always larger than the amount still further recovered in any subsequent hour.[1]

[1] The exceptions have been very few.

That the total amount recovered very seldom exactly equals the total amount previously lost, and is sometimes less and sometimes greater than it. There are consequential changes in the index-error of aneroids.

§ 36. It can be seen from the observations already quoted that there are frequent changes in the initial index-errors of aneroids. If, for example, Col. 4 of the table on p. 30 is compared with Col. 1, it will be noticed that twelve aneroids out of the thirty-six which are there reported upon— those marked by asterisks—recovered, *in a week,* the whole of the amounts which they had previously lost in a week at the mentioned pressures, and gained something more. The twelve all read *higher* at the expiration of *one week* after return to normal pressure than they read before reduction of pressure. At the end of one week they had not completed their 'recovery.' If the whole thirty-six instruments had been watched it would probably have been found that they continued to gain (though in minute quantities) for several weeks longer, and at the expiration of a *month* there would have been about half of them reading *higher* than before reduction of pressure (in comparison with the Mercurial Barometer), and the rest reading *lower* than before reduction of pressure.

§ 37. The considerable length of time over which the 'recovery' extends, and the frequency in alteration of index-errors, will perhaps be better apprehended by inspection of the table on p. 33. In an experiment (which will be referred to in Part 3 [1]) I kept twenty-two aneroids at a pressure of 21·692 inches for a week, and watched the behaviour of the instruments for thirty days after their return to normal pressure. In Col. 2 of the table the errors are stated that these twenty-two aneroids possessed immediately before *reduction* of pressure. In Col. 3 the errors are stated which they were found to have immediately after *restoration* of pressure. In Col. 4 their errors fourteen hours later, and so on. If the eye is carried along each line, from left to right across the page, it will be seen that there is a continual lessening of minus errors and increase in plus ones, showing that recovery was still proceeding after the twelfth day. Then, if the eye is carried *down the columns,* it will be found that immediately after restoration of pressure *one* aneroid already read higher than it did before reduction of pressure. In Col. 4 there are *five* which read higher; in Col. 5 there are *nine,* and in Col. 6 there are *sixteen,* marked by asterisks.

§ 38. In some of the aneroids which have been under examination for years, and have been employed in numerous experiments, I have noticed a continual tendency to augmentation of error in the + direction, whilst in others there has been a continual tendency in the − direction; or, to employ a term which is used in connection with other instruments, some have appeared to have a gaining and others to have a losing rate. It seems to me, also, that a gaining rate not unfrequently changes to a losing one, or the contrary; but upon this point I speak with hesitation, through having been unable to retain instruments for a length of time sufficient to arrive at definite conclusions.

[1] See pp. 37-39.

TABLE SHOWING THE 'RECOVERY' OF 22 ANEROIDS UP TO THE END OF 30 DAYS (AFTER THEY HAD BEEN SUBMITTED TO A PRESSURE OF 21·692 INCHES FOR ONE WEEK), AND THE ALTERATIONS IN THEIR INDEX-ERRORS.

1. Maker's name and mark.	2. Index-error at natural pressure before reduction of pressure.	3. Index-error after pressure was restored.	4. Index-error 14 hours after pressure was restored.	5. Index-error 12 days after pressure was restored.	6. Index-error 30 days after pressure was restored.	
	inches.	inches.	inches.	inches.	inches.	
Hicks . (1)	+ 0·013	− 0·262	− 0·074	− 0·071	+ 0·100	*
„ . (2)	+ 0·033	− 0·202	− 0·044	+ 0·049	+ 0·145	*
„ . (3)	+ 0·103	− 0·067	+ 0·106	+ 0·129	+ 0·170	*
„ . (4)	+ 0·073	− 0·172	− 0·044	− 0·016	+ 0·140	*
„ . (5)	− 0·032	− 0·342	− 0·154	− 0·101	− 0·040	
„ . (6)	+ 0·103	− 0·092	+ 0·046	+ 0·059	+ 0·120	*
„ . (F)	+ 0·248	+ 0·133	+ 0·276	not	observed	
„ . (9)	+ 0·133	+ 0·008	+ 0·136	+ 0·149	+ 0·190	*
Casella . (580)	+ 0·483	+ 0·138	+ 0·376	+ 0·399	+ 0·455	
Hicks . (37)	+ 0·078	− 0·162	+ 0·046	+ 0·129	+ 0·160	*
„ . (38)	− 0·057	− 0·192	− 0·074	− 0·056	+ 0·005	*
„ . (39)	− 0·137	− 0·592	− 0·249	− 0·151	− 0·070	*
„ . (40)	+ 0·478	+ 0·333	+ 0·416	+ 0·349^d	+ 0·470	
„ . (41)	+ 0·343	+ 0·278	+ 0·301	+ 0·259^d	+ 0·360	*
„ . (42)	− 0·007	− 0·122	+ 0·006	+ 0·024	+ 0·075	*
„ . (10)	− 0·177	− 0·517	− 0·274	− 0·231	− 0·160	*
„ . (11)	− 0·407	− 0·607	− 0·499	− 0·441	− 0·450^d	
„ . (12)	− 0·077	− 0·222	− 0·104	− 0·091	− 0·060	*
Hilger . —	+ 0·028	− 0·182	− 0·004	+ 0·034	+ 0·120	*
Hicks . (8)	+ 0·253	+ 0·308	+ 0·466	+ 0·499	+ 0·505	*
Casella . (1021)	− 1·117	− 1·602	− 1·384	− 1·296	− 1·245	
Cooke . (170)	− 0·227	− 0·422	− 0·249	− 0·186	− 0·140	*

^{d} Discordant.

§ 39. These alterations in the index-errors of aneroids are doubtless continually going forward under the natural variations of atmospheric pressure which occur everywhere. From a single fluctuation of, say, an inch, the gain or loss may be so small as to be imperceptible ; but in course of time they mount up and begin to tell, and in this way aneroids acquire (as many persons must have remarked) considerable + or − errors when put away in drawers, and repose, so to speak, out of use.

From the foregoing experimental comparisons of the aneroid against the mercurial barometer in the workshop it will be seen that one may be led into very serious errors in observing variations of pressure by the employment of aneroids. 1. By reason of the loss upon the mercurial which com-

F

mences immediately pressure is reduced, or, 2. through the recovery which sets in immediately pressure is restored ; or, 3. by the changes in index-errors which are continually happening.[1]

I have thought it advisable in the present part of the paper to confine myself as closely as possible simply to comparisons of the indicatious of the two classes of barometer, without regard to the uses to which observations of differences in atmospheric pressure may be applied. The determination of altitudes is amongst the most important of these uses, and in Part 3 I venture to offer a few remarks upon this subject.

---

[1] At the commencement, I employed the apparatus which was in daily use at Mr. Hicks' establishment. It was soon found necessary to duplicate this in order to facilitate progress, and subsequently I set up a similar apparatus in my own workshop. It became necessary also to duplicate this, and finally I had a receiver 30 inches high constructed, in order to have twenty or more aneroids under examination at one time.

Upon going one day to Mr. Hicks' establishment, I found the aneroid department in some confusion, and learned that just before my arrival an accident had happened to an apparatus which we were keeping experimentally at a pressure of 16 inches. With a report like a rifle, plate glass more than a quarter of an inch thick was crushed in by the external pressure, and rebounded in a multitude of fragments over the room. The apparatus was shattered, and an attached mercurial barometer was destroyed.

Mr. Murray, the head of the department, was close by at the time, and said it seemed to him like an explosion, as he only saw the rebounding splinters flying in all directions. As soon as it could be done, Mr. Hicks replaced this apparatus at his own cost, and the experiment was recommenced.

In all the experiments made at Mr. Hicks' establishment, the mercurial barometers were first read by Mr. Murray and the aneroids were read by myself. We then checked each other's readings.

PART 3.

UPON THE USE OF THE ANEROID BAROMETER IN DETERMINATION OF ALTITUDES.

In the second part of this paper it has been shown, 1. that aneroids lose upon the mercurial barometer when submitted to diminished pressure ; 2. that aneroids recover some portion of this loss upon restoration of pressure ; and 3. that there are frequent changes in the index-errors of aneroids. From the most cursory inspection of this part of the paper it will be apparent that an observer in the field, endeavouring to determine altitudes above the sea-level by comparison of aneroid readings against barometric observations at lower stations, may be (and often must be) led into very serious errors through these several causes. In the remarks which follow I shall endeavour to show some ways in which this liability to error may be guarded against.

§ 40. When an aneroid experiences diminished pressure (as it does when a traveller is ascending) it at once commences to lose upon the mercurial barometer, and to indicate a greater diminution of pressure than the truth.[1] The traveller is thus led to overestimate his altitude, no matter whether he attempts to gain an idea of it by mere inspection of the aneroid, or by comparing the aneroid readings with other barometric readings at lower stations. This tendency to lose continues for several weeks. When it ceases (and it will cease in a few weeks at a constant pressure)[2] the aneroid will then indicate the natural (diurnal or hourly) fluctuations in atmospheric pressure almost as accurately as the mercurial barometer itself ; but, *until it ceases, natural (diurnal or hourly) variations in atmospheric pressure cannot be accurately observed from aneroids. It may, indeed, happen that the index of an aneroid may be falling (or indicating diminution in pressure) at a time that pressure is actually increasing.*

§ 41. When an aneroid experiences *restoration of pressure* (after having, previously, experienced diminished pressure) it at once commences to recover, or regain, its previous loss. This recovery almost always goes on for several weeks, and when it ceases (but not until then) the aneroid will again indicate natural (hourly or diurnal) variations with reasonable accuracy.

---

[1] It is assumed in the remarks throughout this part of the paper that aneroids are correctly graduated, and read truly against the mercurial barometer when pressure is reduced *instantaneously*.

[2] And the loss will recommence, and proceed as before, if the aneroid is taken to a greater elevation (that is, experiences still lower pressure).

§ 42. The tendency to lose, and to recover, is most active during the first day of change of pressure, but it is well marked for several days (see table on p. 27). Hence aneroids which may be employed by travellers at upper or lower stations (with the view of applying corrections to simultaneous observations at lower or upper stations) *ought to be installed at such upper or lower stations for several days before being used for simultaneous observations.*

§ 43. If an aneroid should be taken to some lofty position (thereby experiencing diminution of pressure), and should be allowed to remain there for several weeks, it will get into what may be termed " a condition of repose." If the diminution in pressure is large, the instrument will certainly acquire a considerable error ; but at the end of a few weeks this error will remain tolerably constant,[1] so long as the aneroid remains at one level (that is, at a nearly constant pressure) ; and, if it should then be used rapidly, the differences of pressure which will be indicated by it will closely accord with those which would be indicated by a mercurial barometer, as *the error will be carried along the scale.*[2]

§ 44. The fact that aneroids which have acquired large index-errors may nevertheless (under certain conditions) be usefully employed for measurement of differences of level is one of great importance to travellers or surveyors, who must, however, always bear in mind *the best results will be attained when the least possible amount of time intervenes between the readings at two stations upon different levels.*

§ 45. The fact that index-errors are commonly carried along the scale when observations are made with rapidity accounts for some of the remarkable agreements which were observed in Ecuador, and I propose to revert for a few minutes to the instance mentioned in § 12. On the journey to the north of Quito,[3] I carried Merc. Bar. No. 558, and aneroids A and B, and upon arrival at the top of the great ravine of Guallabamba took simultaneous observations of the mercurial and aneroids. At the bottom of the ravine, two hours and a half later, readings of all three were repeated with the following result :—

Date.	Barometer.	Read at top.	Read at bottom.
Mar. 27, 1880	Merc. bar. 558 (red. to 32° Faht.)	21·692	23·929
do.	Aneroid A	21·140	23·400
do.	do. B	19·940	22·200

The rise of the Mercurial Barometer was 2·237 inches.

do.	do.	Aneroid A	„	2·260 „
do.	do.	„ B	„	2·260 „

---

[1] But not *perfectly* so. It has been pointed out in § 39 that long-continued diurnal fluctuations in pressure (although small) may in course of time produce marked changes in index-errors.

[2] This will be illustrated experimentally a few pages later.

[3] See *Travels amongst the Great Andes of the Equator*, chap. xii, and § 12 of this paper.

In this instance, two aneroids, one of them having an index-error of more than half an inch, and the other having an index-error of about one inch and three-quarters, were employed for a measurement of difference of level. The two aneroids indicated precisely the same difference of *pressure* between the upper and lower stations, and differed only *one per cent* from the difference of pressure indicated by the mercurial. How are these facts to be accounted for ?

They appear to me to be explicable in the following manner. The large index-errors were the result of the invariable " loss upon the mercurial " which takes place in aneroids upon sustaining diminished pressure, added to the change in index-error which frequently accumulates when aneroids are subjected to large and frequent *fluctuations* of pressure. Prior to March 27, the two aneroids had remained for more than four weeks at Quito at a nearly constant pressure of 21·600 inches, and had thus got into " a state of repose." Their index-errors were remaining *constant*, and the instruments were affected by the small diurnal variations in pressure, which occurred at Quito, in the same way as the mercurial barometers. If upon arriving at the edge of the ravine they could have been transported *instantaneously* to the bottom of it, the difference in pressure which they would have indicated would, I do not doubt, have almost exactly coincided with that indicated by the mercurial barometer. The fact was they indicated a difference in pressure 0·023 of an inch more than the mercurial, and this excess I attribute to the tendency to ' recover ' which sets in immediately upon restoration of pressure.

This experience left a strong impression upon my mind. I have repeated the experiment in the workshop with the following results.

§ 46. I took twenty-two aneroids (purposely including amongst them those having the largest index-errors that could be found amongst the instruments which were available) and reduced them to a pressure of 21·692 inches (the pressure at the top of the ravine of Guallabamba). I kept them constantly at this pressure for one week (so that the greater part of the " loss upon the mercurial " might occur),[1] and then *gradually* restored pressure during two and a half hours, until it stood at 23·929 inches (the pressure at the bottom of the ravine of Guallabamba). In Column 1 of the annexed table the readings are given which were taken before reduction of pressure, and in Col. 2 the various errors of the aneroids at that time. In Col. 3 the readings of the aneroids are given which were taken one week after they had been reduced to 21·692 inches, and in Col. 4 the various errors of the aneroids at that time. In Col. 5 the readings are given which were taken when the mercurial barometer indicated a pressure of 23·929 inches, and in Col. 6 the rise is stated which was indicated by all the barometers (the differences between the readings in Cols. 3 and 5). In Cols. 7 and 8 the readings and errors are given upon return to natural pressure.

---

[1] To have repeated the conditions of the experiment in the field, these aneroids should have been kept at 21·692 inches for four or more weeks. There were other demands upon my apparatus at the time, which prevented me from doing so.

REPETITION IN THE WORKSHOP OF "THE GUALLABAMBA EXPERIMENT."

Description of instrument.	1. At natural pressure. Reading.	2. Error.	3. Pressure having been reduced for one week. Reading.	4. Error.	5. Pressure partially restored during 2½ hours. Reading.	6. Rise Indicated.	7. On return to natural pressure. Reading.	8. Error.
	Inches.	Inches.	Inches.	Inches.	Inches.	Inches.	Inches.	Inches.
Attached merc. bar. reduced to 32° Faht.	29·697	…	21·692	…	23·929	2·237	29·842	…
Aneroid 1 (Hicks)	29·710	+0·013	21·425	−0·267	23·580	2·155	29·580	−0·262
„ 2 „	29·730	+0·033	21·425	−0·267	23·580	2·155	29·640	−0·202
„ 3 „	29·800	+0·103	21·425	−0·267	23·680	2·255	29·775	−0·067
„ 4 „	29·770	+0·073	21·425	−0·267	23·625	2·200	29·670	−0·172
„ 5 „	29·665	−0·032	21·200	−0·492	23·420	2·220	29·500	−0·342
„ 6 „	29·800	+0·103	21·470	−0·222	23·640	2·170	29·750	−0·092
„ F „	29·945	+0·248	21·350	−0·342	23·710	2·360	29·975	+0·133
„ 9 „	29·830	+0·133	21·530	−0·162	23·735	2·205	29·850	+0·008
„ 580 (Casella)	30·180	+0·483	21·580	−0·112	23·840	2·260	29·980	+0·138
„ 37 (Hicks)	29·775	+0·078	21·200	−0·492	23·375	2·175	29·680	−0·162
„ 38 „	29·620	−0·057	21·275	−0·417	23·520	2·245	29·650	−0·192
„ 39 „	29·560	−0·137	20·569	−1·123	22·850	2·281	29·250	−0·592
„ 40 „	30·175	+0·478	21·850	+0·158	24·125	2·275	30·175	+0·333
„ 41 „	30·040	+0·343	21·820	+0·128	24·050	2·230	30·120	+0·278
„ 42 „	29·690	−0·007	21·355	−0·337	23·600	2·245	29·720	−0·122
„ 10 „	29·520	−0·177	20·850	−0·842	23·050	2·200	29·325	−0·517
„ 11 „	29·290	−0·407	20·875	−0·817	23·175	2·300	29·235	−0·607
„ 12 „	29·620	−0·077	21·240	−0·452	23·500	2·260	29·620	−0·222
„ — (Hilger)	29·725	+0·028	21·870	+0·178*	23·900	2·130	29·660	−0·182
„ 8 (Hicks)	29·950	+0·253	21·950	+0·258*	24·200	2·250	30·150	+0·308
„ 1021 (Casella)	28·580	−1·117	20·030	−1·662	21·980	1·950	28·240	−1·602
„ 170 (Cooke)	29·470	−0·227	21·250	−0·442	23·535	2·285	29·420	−0·422

* Discordant.

The immediate object of this experiment was to see how repetition in the workshop would accord with the experience in the field. It will be found that it agreed in a remarkable manner. The mean of the whole of the aneroid differences in Col. 6 is 2·218 inches; or, if the reading of No. 1021 (Casella) is rejected,[1] and the mean is taken of the twenty-one others, it will be found to be 2·231 inches,—a difference of only six-thousandths of an inch from the rise in pressure indicated by the mercurial barometer. If the aneroid observations are taken singly, it will be found that six of them differ from the mercurial barometer only to the extent of 0·018 of an inch or less, and that the extremes (rejecting No. 1021) are 2·130 and 2·360 inches.

From this experiment (as well as from many others) it appears that although aneroids under diminished pressure acquire considerable index-errors, and often acquire huge ones under a long continuance of largely diminished pressure combined with frequent fluctuations in pressure, they still continue to march inch by inch with the mercurial barometer; and upon a single operation involving a difference of pressure even of several inches may come very near to the truth, provided the operation is quickly performed.

§ 47. The correct measurement of differences of *pressure* is, however, only a step in the barometric determination of differences of *level*. Every inch of the barometer has its own special value in calculations of altitude. The inch embraced between 31 and 30 inches is worth about 894 feet, while that between 15 and 14 inches is worth about 1880 feet. In order to ascertain differences of level correctly it is necessary for an observer to know not only the difference of pressure but also the particular inches embracing his difference of pressure. The value in English feet of the difference between 19·940 and 22·200 inches (the readings of aneroid B at the ravine of Guallabamba) is ten per cent greater than the value of the difference between 21·692 and 23·929,—the readings of the mercurial barometer.[2] *Even for measurement of differences of level, it is highly desirable that a traveller should, at all times, have the means of ascertaining the index-errors of his aneroids.*

§ 48. The importance of being able to do this is far greater in observations for determination of elevation above the level of the sea. The mistake which the index-error of an aneroid then introduces into calculations is the whole value of the difference of its readings at the upper station from the true barometric reading.[3] Aneroid B, at the top of the ravine of Gualla-

---

[1] As it fairly may be. This is an old aneroid, which, prior to this experiment, had been reposing in a drawer for about eighteen years, and did not work freely.

[2] For the purposes of this passage, and for all the computations which have been necessary throughout this paper of the value of the barometric inch at different pressures, I have used the Table prepared at the Royal Observatory, Greenwich, entitled *Corresponding Numbers of Elevation in English Feet, and of Readings of Aneroid or Corrected Barometer in English Inches; the Mean of Atmospheric Temperatures being 50° Fahrenheit.* See pp. 56, 57.

[3] See the last columns of the tables upon pp. 21-24.

bamba, read 19·940 and the mercurial barometer 21·692 inches. The difference of these two readings is 1·752 inches, and the value of this, at the elevation in question, is above 2000 feet. Although this is a gross case, more extreme ones are possible.

§ 49. I desire to lay especial stress upon the changes which occur in the index-errors of aneroids. From prolonged observation, it appears to me that they seldom remain stationary, and are never permanent. They take place from a variety of causes. In the course of my experiments in the workshop I have seen the index-errors of an aneroid grow to as much as four inches; in several instances there have been alterations of more than an inch; and in numerous instruments there have been alterations from scarcely appreciable errors to + or − errors of two to four-tenths of an inch.[1] Even the least of these amounts (0·200 of an inch) is of consequence in the calculation of altitudes, and it has a considerable value when great heights (low pressures) are concerned.

The continual variations which occur in the index-errors of aneroids have not, I think, hitherto been sufficiently recognized. Index-errors have generally been treated as if they remained *constant;* and, to correct them, it has been considered sufficient to apply as constant corrections (perhaps for the space of *years*) the amounts stated in "certificates of examination." These certificates [2] have a very limited value. They point out, no doubt correctly, the errors which were exhibited *at the time of examination;* and, if the certified aneroids are employed speedily after that date, and again experience reductions in pressure *of the same amount, and for the same length of time* as during the examination, it may be useful to pay attention to the errors which can be deduced from the certificates. A single week, however, of active use (that is to say, of considerable variations in pressure) may considerably alter the index-errors, and will certainly change them to an appreciable extent, while under more prolonged use further variations will most likely occur.

It seems to me, therefore, *indispensable for all those who aim at correct measurement of altitude above the level of the sea, by means of aneroids, and especially for those engaged upon prolonged journeys, to be at all times able to determine their index-errors.*

§ 50. I now draw attention to a feature in the working of aneroids which has as yet only been casually referred to in this paper, namely, the want of correspondence between differences of pressure indicated by them in *ascending* and in *descending* observations. Under certain conditions descending readings indicate less difference of pressure than ascending ones, and under other conditions the reverse may be the case. These facts can be utilized in deter-

---

[1] The index-errors stated in Col. 2 of the table upon p. 38, with the exception of F (Hicks), 580 and 1021 (Casella), and 170 (Cooke), were all acquired in the course of the experiments which are described in Part 2. One of them, it will be seen, is as much as + 0·478, and another − 0·407 of an inch. These alterations occurred though the instruments were used with the utmost tenderness. Under the rougher usage to which they would necessarily be submitted in the field there would be liability to still greater changes.

[2] In their present form. See pp. 51-54.

mination of altitudes. I shall presently endeavour to show experimentally the effect which is produced by the intervention of a greater or less space of time ; but, before doing so, I will restate a few of my facts in a somewhat different manner.

§ 51. All aneroids commence to lose upon the mercurial barometer directly pressure is reduced. They lose a certain amount if pressure is reduced to the extent of an inch, they lose a larger amount if it is reduced 2 inches, and more and more for each successive inch. Also the loss is greater in a day than it is in an hour, greater in a week than it is in a day, and it continues to augment perceptibly for about a month. Further, immediately pressure commences to be restored, an aneroid endeavours to recover the previous loss. This latter tendency is more and more important as each inch of pressure is restored, and after return to normal pressure it continues to operate (though with constantly diminishing force) for several weeks.

§ 52. It must therefore be apparent that the difference of pressure (or difference of level) which will be indicated by an aneroid between any two stations will be influenced by the greater or less amount of time which may intervene between the readings at the two stations. In the case of a mountain 5000 feet high (or any other height) one result will be attained from an aneroid if there is an interval of four hours between a reading at the bottom and another at the top, a different result will be attained if there is an interval of twelve hours, and yet another if there is an interval of twenty-four hours ; and the difference of pressure (or the altitude) which will be indicated will be greater in the second case than in the first, and greater still in the third than in the second. It will not, perhaps, be so readily apparent why descending readings, *closely following upon ascending ones,* always differ from them, and I will state three imaginary cases in which this will occur.

§ 53. Let us imagine the case of an ascent being made in five hours, of a mountain having a pressure of 25 inches reigning at its top and a pressure of 30 inches at its base. The diminution in pressure which will occur during the ascent will be gradual, and the effect which will be produced on the aneroid by the time it arrives at the summit will be much less than the effect which would have been produced upon it if it had sustained a *constant* diminution in pressure of 5 inches for five consecutive hours. The loss upon the mercurial barometer will nevertheless be well marked, and at the end of the fifth hour, as the aneroid will then experience a *constant* reduction in pressure of 5 inches, it will lose rapidly. If a reading is taken on the summit immediately upon arrival there, the altitude observed by inspection of the aneroid will be somewhat (but not much) *greater* than the truth,[1] and it will be *less* than the altitude which will be observed if another reading is taken, an hour later. The result in

---

[1] It is of course assumed that the aneroid has been correctly graduated, to read instantaneously against the mercurial barometer ; and it is also assumed that there is no change in atmospheric pressure, within the period involved, either at the higher or lower station.

each case is due to the tendency to lose upon the mercurial barometer, —which will still be actively at work at the end of the sixth hour, and at the end of that time let us suppose the descent is commenced.

Immediately the descent is commenced pressure begins to be restored. The tendency *to lose* still continues, though with constantly diminishing force. For it is less potent at a pressure of 26 inches than it was upon the summit at a pressure of 25 inches, less potent still at 27 inches, and it dies out on return to normal pressure. But the moment pressure begins to be restored (as it does directly the descent commences) the tendency to *recover* sets in. This is feeble at first (when the restoration in pressure is slight) and it becomes stronger and stronger as normal pressure is approached. Thus, during the descent, there are two tendencies at work—the tendency to lose in consequence of diminution in pressure, and to recover in consequence of restoration of pressure. These counteract each other, and *the effect is the recovery is retarded.* I have found experimentally that, through the tendency to recover, a greater difference of pressure (or greater altitude) than the truth would always be obtained from descending observations (following ascending ones) if this were not checked by the tendency to lose ; and I have also found experimentally that retardation of the recovery causes descending readings to indicate less difference of pressure than ascending ones.[1]

§ 54. In Case 2, let us suppose the same mountain is ascended again, at the same rate, and that the traveller remains an entire *day* on the summit. In this case, if the traveller reads his aneroid immediately upon arrival at the summit he will, as before, obtain an altitude which will be somewhat greater than the truth ; and should he read it again at the expiration of twenty-four hours he will obtain an altitude considerably higher than he did from his second reading upon his first ascent, inasmuch as the aneroid will have been subjected continuously to a diminution in pressure of 5 inches *for an entire day (instead of a single hour)*, and will have been losing upon the mercurial barometer during the whole time. At the end of this time, the tendency to lose will have become much less active (see table on p. 27) ; and, should he now commence to descend, the tendency of the aneroid to recover will be less checked than before. In consequence of this, over precisely the same ground, even if he comes down at precisely the same rate as upon his first descent, his descending readings will yield a result differing from that obtained on his first descent, and will indicate a *greater* difference of pressure, and greater difference of level between the summit and the base than upon the first occasion.

§ 55. For Case 3, let us suppose the same mountain is ascended again, at the same rate as before, and that the traveller remains an entire *month* on

---

[1] Not, however, necessarily less than the truth. See later. Sometimes the opposition of these tendencies produces a perfect correction ; and the aneroid, at one point of its scale, will indicate the same differences as the mercurial barometer. See the Kew certificate given on p. 51.

the summit. At the end of this time the tendency to lose will have ceased (or will be exerted very feebly). Thus, when the descent is commenced, the tendency to recover will be unretarded ; and, in consequence of this, over precisely the same ground, and during the same lapse of time as before, the descending readings will yield a result which will differ from those obtained upon the first two descents, and will indicate a *greater* difference of pressure, and *greater* difference of level than upon the first *two* occasions.[1]

§ 56. The remarks which have been made upon pp. 41-2 will probably be better apprehended after the tables have been examined that accompany the following experiments. These two experiments are designed to show that, in measuring precisely the same amount of difference of pressure upon several occasions, aneroids will yield dissimilar results upon each occasion, if the time conditions are dissimilar ; and also to exhibit the manner in which exact knowledge of the behaviour of aneroids may be turned to account in the measurement of pressures, and through them of altitudes.

§ 57. EXAMPLE 1. The case is supposed of a traveller, provided with aneroids, starting from the level of the sea to ascend a mountain—the height of which is unknown. It is supposed that no change in atmospheric pressure occurs during the expedition, and that the traveller proposes to deduce the altitude of his mountain from means of ascending and descending observations of the differences of pressure indicated by his aneroids.

He is supposed to occupy four hours upon the ascent ; to remain one hour upon the top ; and to descend from the summit to his starting-point (at the level of the sea) in two hours more. To read his aneroids just before his departure from the level of the sea ; upon the summit, one hour after his arrival there ; at the level of the sea immediately upon his return, and again at the level of the sea twenty-two hours later. What error, expressed in barometric inches, is he likely to fall into from his employment of aneroids in this manner ?

For this experiment I used twelve aneroids, all of the watch size.[2] The aneroids were read ; pressure was then reduced *gradually* for four successive hours, at the rate 1·146 inches per hour ; it was maintained at 25·061 inches for one hour, and the aneroids were again read at the expiration of this (the fifth) hour. Pressure was then *gradually* restored during two hours. By the end of that time the aneroids arrived again at the pressure from which they

---

[1] It is scarcely necessary to say that should an aneroid be used for ascending observations, immediately (or shortly) after it has experienced *augmentation* of pressure, the tendency to recover will retard the tendency to lose in precisely the same manner as the tendency to lose retarded the tendency to recover in the instances imagined in §§ 53-55. Thus, if a traveller ascends a mountain upon two successive days, using the same aneroids on each occasion, and starts upon his second ascent (say) ten hours after his return to camp from the first one, he will from his second set of ascending readings obtain a less difference of pressure, and a smaller difference of level, than he did upon his first ascent.

It may be pointed out here that the embarrassment caused by the 'recovery' has greatly protracted these investigations. In the case of repetitions, it has sometimes been necessary to put instruments aside for months.

[2] Because this is the size most commonly used by travellers.

EXAMPLE 1.

Nature of Barometer.	1. Readings at assumed sea-level.	2. Readings on supposed summit, 1 hour after arrival.	3. Showing a fall of	4. Readings on return to sea-level.	5. Showing a rise of	6. Readings 22 hours after return to sea-level.	7. Showing a rise of
	Inches.	inches.	Inches.	inches.	inches.	inches.	inches.
Attached Mercurial Barometer (reduced to 32° Faht.)	29·645	25·061	4·584	29·645	4·584	29·645	4·584
Hicks' Aneroid, No. 29	29·150	24·550	4·600	29·092	4·542	29·146	4·596
do. do. 30	29·805	25·200	4·605	29·772	4·572	29·796	4·596
do. do. 31	29·670	25·020	4·650	29·622	4·602	29·676	4·656
do. do. 32	29·700	25·150	4·550	29·632	4·482	29·666	4·516
do. do. 33	29·810	25·290	4·520	29·762	4·472	29·776	4·486
do. do. 35	29·750	25·105	4·645	29·707	4·602	29·756	4·651
do. do. 37	29·800	25·210	4·590	29·747	4·537	29·786	4·576
do. do. 38	29·690	25·025	4·665	29·632	4·607	29·671	4·646
do. do. 39	29·690	25·025	4·615	29·612	4·587	29·691	4·666
do. do. 40	30·150	25·560	4·590	30·097	4·537	30·151	4·591
do. do. 41	29·975	25·425	4·550	29·932	4·507	29·951	4·526
do. do. 42	29·710	25·075	4·635	29·672	4·597	29·716	4·641
Means of the Aneroid observations			4·605		4·554		4·596
Differences of the means of the Aneroids from the Mercurial Barometer			+0·021		-0·030		+0·012

had started, namely, 29·645 inches. The instruments were then again read, and were again read twenty-two hours later.

In Col. 1 of the annexed table, the readings are given which were taken at the assumed sea-level ; in Col. 2, the readings upon the supposed summit, one hour after arrival ; in Col. 4, the readings immediately upon return to the starting-point ; and, in Col. 6, the readings which were taken twenty-two hours later. In Col. 3 the differences are given between the readings in Cols. 1 and 2 ; in Col. 5, the differences between the readings in Cols. 2 and 4 ; and in Col. 7, the differences between the readings in Cols. 2 and 6. At the bottom of the table I give the "Means of the Aneroids," and the differences of those means from the difference of pressure indicated by the mercurial barometer.

For the present, I invite attention only to the means of the aneroid observations in Columns 3, 5, and 7. The first will be found to be 0·021 of an inch *in excess* of, the second to be 0·030 of an inch *less* than, and the third to be 0·012 of an inch in excess of the difference of pressure indicated by the mercurial barometer.

§ 58. EXAMPLE 2. It is supposed that our traveller's observations were considered unsatisfactory ; that it was said his work was too hurriedly performed, and that he should have taken at least two readings upon the summit, several hours apart, and should have used the means of these two readings.

He is supposed to have made another expedition ; to have started on the first day at 10 a.m., and to have reached the summit at 5 p.m. ; to have remained there until 5 p.m. on the second day ; and then, as before, to have descended in two hours to the level of the sea. He is supposed to have read his aneroids just before departure from the level of the sea ; to have read them again at 11 a.m. and at 5 p.m. on the day which was passed on the summit ; and again immediately on return to the level of the sea. What superiority would these observations possess over the former ones ?

In this experiment I used the same aneroids as before, and read them against the mercurial barometer at the hours mentioned above. During the seven hours which are supposed to have been occupied by the *ascent*, pressure was reduced at the rate of 0·655 of an inch per hour, and during the descent it was restored gradually.

In Col. 1 of the table (Example 2) the readings are given which were taken at the assumed sea-level ; in Col. 2 the readings on the supposed summit, eighteen hours after arrival ; in Col. 4 the readings on the supposed summit twenty-four hours after arrival ; and in Col. 6 the readings upon return to the starting-point. In Col. 3 the differences are given between the readings in Cols. 1 and 2 ; in Col. 5 the differences between Cols. 1 and 4 ; and in Col. 7 the differences between Cols. 4 and 6. I again ask attention first of all to be given to the means of the aneroid observations in Columns 3, 5, and 7. The first will be found to be 0·111, the second to be 0·134, and the third to be 0·073 of an inch in excess of the true difference of pressure observed from the mercurial barometer.

EXAMPLE 2.

Nature of Barometer.	1. Readings at assumed sea-level.	2. Readings on supposed summit, 18 hours after arrival.	3. Showing a fall of.	4. Readings on supposed summit, 24 hours after arrival.	5. Showing a fall of.	6. Readings on return to sea-level.	7. Showing a rise of.
	inches.	Inches.	Inches.	inches.	inches.	inches.	inches.
Attached Mercurial Barometer (reduced to 32° Faht.)	29·709	25·125	4·584	25·125	4·584	29·709	4·584
Hicks' Aneroid, No. 29	29·230	24·550	4·680	24·540	4·690	29·165	4·625
do. „ 30	29·860	25·200	4·660	25·210*	4·650	29·825	4·615
do. „ 31	29·740	25·040	4·700	25·010	4·730	29·670	4·660
do. „ 32	29·775	25·140	4·635	25·130	4·645	29·685	4·555
do. „ 33	29·870	25·280	4·590	25·260	4·610	29·845	4·585
do. „ 35	29·840	25·035	4·805	25·015	4·825	29·775	4·760
do. „ 37	29·860	25·015	4·845	24·990	4·870	29·745	4·755
do. „ 38	29·690	25·000	4·690	24·965	4·725	29·655	4·690
do. „ 39	29·760	24·875	4·885	24·840	4·920	29·625	4·785
do. „ 40	30·200	25·600	4·600	25·555	4·645	30·160	4·605
do. „ 41	30·040	25·450	4·590	25·415	4·625	30·010	4·595
do. „ 42	29·750	25·090	4·660	25·065	4·685	29·715	4·650
Means of the Aneroid observations	...	...	4·695	...	4·718	...	4·657
Differences of the means of the Aneroids from the Mercurial Barometer	...	...	+0·111		+0·134		+0·073

* A discordant reading.

§ 59. We have thus obtained six different results in the measurement of a difference of pressure of 4·584 inches. One is 0·030 of an inch *less* than the truth, and the others are 0·012, 0·021, 0·073, 0·111, and 0·134 of an inch *greater* than the truth. Some may say "*this is another instance of the uncertainty of aneroids.*" Internal examination of these readings will, however, show that aneroids work with great certainty, and follow rules which can be defined with some precision. Let us first examine the ascending readings.

After having had pressure reduced gradually in *four* hours to the extent of 4·584 inches, and after being kept at that pressure for *one* hour, the mean of the twelve aneroids indicated a fall in pressure 0·021 of an inch more than the truth.

After having had pressure reduced gradually in *seven* hours to the extent of 4·584 inches, and after being kept at that pressure for *eighteen* hours, the mean of the twelve aneroids indicated a fall in pressure 0·111 of an inch more than the truth.

After being kept at this reduction of pressure for *six hours more*, the mean of the twelve aneroids indicated a fall in pressure 0·134 more than the truth.

This behaviour is consistent with the rule which has been already stated, namely, that all aneroids lose upon the mercurial barometer upon being submitted to diminished pressure, and that *the extent of the loss in any operation will depend upon the lapse of time as well as upon the extent of the diminution in pressure.*

The behaviour of the individual instruments in these experiments will also be found consistent. If Col. 3 of tab. Example 1 is compared with Col. 3 of tab. Ex. 2, it will be found that every single instrument in the latter column indicated a greater fall than in the former; and if Cols. 3 and 5 of tab. Ex. 2 are compared, it will be found that the six additional hours upon the supposed summit produced a distinct increase in the loss upon the mercurial in eleven out of the twelve instruments.[1]

§ 60. Turning now to the means of the *descending* readings we find the first giving a result 0·030 of an inch *less* than the truth, and the others 0·012 and 0·073 of an inch *more* than the truth. The time conditions were different in each of these cases, and in this fact I seek for an explanation.

The first of these means (– 0·030) is derived from the readings of twelve aneroids, which in *four* hours had pressure reduced to the extent of 4·584 inches, were then kept at that pressure for *one* hour, and then had pressure restored gradually during two hours. The resulting mean is consistent with the general behaviour of aneroids. It is the rule that

---

[1] The reading of aneroid 30 in Col. 4 of tab. Ex. 2 is marked 'discordant.'

When an aneroid is under the receiver of an air-pump, it is not always possible to assist the movements of the index-needle by tapping. The discrepancy between the two readings of No. 30 may have occurred through some slight impediment which would have been immediately overcome if the instrument could have been tapped in the usual way.

all descending observations of aneroid barometer which follow ascending ones indicate less increase in pressure than the truth if they are made quickly (i.e. *if only a short time elapses between the readings at the higher and the lower station*), *provided also that the ascending observations have been made quickly* (*that is, that only a short time has elapsed between the readings at the lower and the upper station*).[1]

The second of the means of the descending observations ($+0·012$) is derived from readings of the same aneroids, twenty-two hours after they had returned to normal pressure, and is consistent with the general behaviour of aneroids. The rule is that *all aneroids without exception lose upon the mercurial barometer when submitted to diminished pressure, and recover a portion of the previous loss when pressure is restored.*[2] In this instance they recovered $0·042$ of an inch (the difference between $-0·030$ and $+0·012$) in twenty-two hours.

The third of the means of the descending observations ($+0·073$) is derived from the readings of the same aneroids after they had experienced reduction of pressure to the extent of $4·584$ inches in *seven* hours, and had then been kept continuously at that pressure for *twenty-four* hours. At the end of that time they had pressure restored, as before, in the course of two hours. The result is not discordant although it differs from the others. The instruments experienced reduction of pressure for a greater length of time than before, and consequently lost more than before. At the end of *twenty-four* hours the tendency to lose was much less active than it was on the first occasion (at the end of *one* hour). The increase in the difference of pressure that they indicated (namely $4·657$ inches, as compared with $4·554$ and $4·595$) was due to there being a larger amount to recover, and to the recovery being less retarded than before by the tendency to lose (see §§ 53–5).

The behaviour of the individual instruments is consistent in the descending readings. If Cols. 5 and 7 of tab. Ex. 1 are compared, it will be seen that every single instrument indicated a greater difference of pressure in the latter than in the former; and the same will again be found if Col. 7 of tab. Ex. 1 is compared with Col. 7 of tab. Ex. 2.

§ 61. If the means of the aneroids are now examined to see what combination gives the closest approximation to the mercurial barometer, it will be found that the best result is obtained by taking the mean of the means of Columns 3 and 5, in Example 1.

Mean of fall in pressure indicated by aneroids (Col. 3) . . 4·6050 inches.
    do.   rise   do.   do. (Col. 5) . . 4·5540 „

Mean of the rise and fall in pressure indicated by aneroids . 4·5795 „
    do.   do.   do.   Merc. Bar. 4·5840 „

Error of the aneroids, $-0·0045$ of an inch.

---

[1] See § 53, and the Kew certificate upon p. 51.
[2] See § 35, p. 31, of this paper.

It will be found that no result so accurate can be obtained from any combination of the means in Example 2, and that *the inferiority of the latter is distinctly traceable to the greater lapse of time between the observations.*

It is possible, therefore, *if aneroids are used in this way,* in measurement of a difference of level involving a difference of pressure of 4·584 inches to get a result only 0·0045 of an inch from the truth. Critics may say that this close result is entirely accidental ; or that no traveller would employ twelve aneroids upon such an expedition ; or that one would be as likely to be provided with several of the worst as with several of the best of the dozen which were used in the experiments. Let us therefore examine some of the results separately.

The aneroid which yielded the worst result was No. 33.

No. 33, ascending observation, 4·520 ⎫
   do. descending   do.   4·472 ⎬ Mean 4·4960.

The two aneroids which yielded the worst combined mean result were Nos. 32 and 33.

No. 32, ascending observation, 4·550 ⎫
   do. descending   do.   4·482 ⎪
No. 33, ascending   do.   4·520 ⎬ Mean 4·5060.
   do. descending   do.   4·472 ⎭

The two aneroids with the largest initial index-errors were Nos. 29 (− 0·495) and 40 (+ 0·505).

No. 29, ascending observation, 4·600 ⎫
   do. descending   do.   4·542 ⎪
No. 40, ascending   do.   4·590 ⎬ Mean 4·5672.
   do. descending   do.   4·537 ⎭

The aneroid which gave the best result was No. 30.

No. 30, ascending observation, 4·605 ⎫
   do. descending   do.   4·572 ⎬ Mean 4·5885.

Error of aneroid No. 30, + 0·0045 of an inch.

Thus, the worst result which was yielded by any single aneroid was less than two per cent in error, and the best one (like the mean of the whole) differed only 0·0045 of an inch from the mercurial barometer.

Twelve aneroids and no more were employed in this experiment, and none of these were specially selected for it. They came from the stock of Mr. Hicks, and the maker was unaware of the nature of the tests to which they would be put. At the commencement of the experiment, care was taken that they were in—what I have ventured to term—"a state of repose," and no such results could have been obtained if this had been neglected.

The reader will by this time perceive that not only are there various ways in which aneroids can be used but that the precise manner in which they can be most advantageously employed will be determined by the necessities of the case. I have first set before him some of the principal causes of error, and have then endeavoured to point out some ways in which these can be guarded against. He has seen that precautions are *desirable* in some instances, and are *indispensable* in others. He has seen that aneroids are at present constructed to read instantaneously, or nearly so, against the mercurial barometer, and he has been shown that they are very materially affected by variations in pressure, that the extent to which they are affected depends upon the greater or less lapse of time, and that they are *unequally* affected by these causes.

§ 62. Some may wish to measure a mound in a few minutes or to bound in balloon ten thousand feet in an hour, and others to spend a day upon the ascent of a mountain or to pass weeks, months, or years at great elevations in the interior of continents. It is impossible to dictate the right course of procedure for every case that may occur. The best results will be attained by those who make themselves acquainted with the rules that govern aneroids in general, and with the particular behaviour of the instruments which may be employed. I offer the following remarks as hints rather than as directions.

1. When only a few *minutes* elapse between readings, such errors as may occur will most likely be due to errors of graduation.

2. When readings are taken a few *hours* apart (for difference of level) it will be advantageous to employ the means of ascending and descending readings (§§ 52, 61).

3. When *days* elapse between readings taken to obtain difference of level, it will be *necessary* to know the amounts which will probably be lost or recovered at the pressures experienced, during the length of time concerned ; and *desirable* to be able to determine index-errors (§ 47).

4. It seems to me *indispensable* for all those who aim at correct measurement of altitudes above the level of the sea by means of aneroids to be at all times able to determine their index-errors.

No one, I think, will regret paying attention to the following recommendations :—1. Prefer open scales.   2. Prefer aneroids of large diameter to the watch size.   3. Prefer aneroids which have been made some length of time.[1]   4. Avoid working aneroids up to the extreme inferior limit of their scales.   5. Treat an aneroid with the same care as a watch.   6. Trust more to the scale of inches than to the scale of feet.

Something may be gained by paying attention to the foregoing hints and recommendations, but the two essential requirements for those who use aneroids in the field upon prolonged journeys are 1. knowledge of the amounts which will probably be lost and recovered by their instruments, and 2. ability to determine their index-errors at any time.

§ 63. I do not venture to suggest what form of certificate would be best adapted for general use.   The settlement of this may well be left

---

[1] Old aneroids (that is, aneroids which have been made for a number of years) are generally more sober and regular in their conduct than young ones. This fact is, I believe, already well known to instrument makers.

CERTIFICATE OF EXAMINATION
ISSUED BY

THE KEW OBSERVATORY,
RICHMOND, SURREY.

ANEROID BAROMETER No. 899.
by   CARY,   London.

Compared with the Standard Barometer of the Kew Observatory (reduced to 32° Fah^(t.)) with the following results.

Pressure.	Correction to Aneroid with a pressure diminishing.	Correction to Aneroid with a pressure increasing.
At 30 in	0.00	+0.20 *
29	−0.05	+0.25
28	−0.05	+0.20
27	0.00	+0.25
26	0.00	+0.30
25	−0.05	+0.30
24	−0.05	+0.30
23	0.00	+0.35
22	0.00	+0.35
21	0.00	+0.35
20	+0.05	+0.35
19	+0.05	+0.30
18	+0.05	+0.25
17	+0.05	+0.20
16	+0.05	+0.15
15	+0.05	+0.10

* It is probable that this correction will have become reduced to that first obtained after the instrument has remained at the normal atmospheric pressure for a short time.

Note.—When the sign of the Correction is +, the quantity is to be *added* to the observed scale reading, and when − to be *subtracted* from it.

G. M. WHIPPLE,
KEW OBSERVATORY, *October* 1888.   SUPERINTENDENT.

to those whose business it is to attend to such matters ; but, whatever
it may be, it should, 1 think, very materially differ from the certificate
given overleaf, which is one recently issued by Kew Observatory. It is
convenient to refer to this certificate for illustration of several of the
points which have been touched upon.

The aneroid No. 899, Cary, has twice passed through my hands—
on the first occasion in 1889, seven months after it had been tested at
Kew, and on the second in 1890, after a journey in the Caucasus upon
which it had been employed. In 1889, when it first came to me,
I tested it under the air-pump in the ordinary manner, reducing pressure inch by inch, and reading it against the attached mercurial barometer at each successive inch.[1] The time occupied in reducing it from
30 to 15 inches amounted to twenty-five minutes. The results of
my comparisons agreed closely, at every inch of the scale, with those
obtained at Kew. At 29 inches I found the aneroid read 0·073 of an
inch too high (the corresponding result at Kew was +0·050), and at
15 inches it read 0·045 too low (Kew result −0·050). As the Kew
certificate only goes to the nearest half-tenth of an inch, their results
and mine may be considered to agree, and to be satisfactory, as they
show that No. 899, Cary, upon being tested independently upon two
occasions, seven months apart, behaved almost alike on the two occasions,
and that the prognostication contained in the passage at the foot of the
Kew certificate was correct.[2]

I then reduced No. 899, Cary, to a pressure of 16 inches, and
kept it at that pressure for eleven days. At the end of a week this
aneroid lost 0·625 of an inch, and at the end of the eleventh day
0·736 of an inch upon the mercurial barometer (the Kew result was
0·050). Thus, if the owner of this aneroid had remained with it at a
pressure of 16 inches for a week, and had employed the correction
stated upon the Kew certificate, he would probably have under-estimated
the pressure to the extent of 0·575 of an inch, which would have
introduced a very large error into computations of altitude. He would
have employed the error due to a diminution of pressure for the space
of about twenty-five minutes instead of that arising in a week, and from
this, I think, it sufficiently appears that it is desirable to present certificates of examination in a form different from that which is given upon
p. 51.

§ 64. The readings of No. 899, Cary, which were taken during the Kew
examination can be recovered by applying the corrections stated upon the
certificate, and are given below. In Col. 1 we have the readings of the Kew
Standard Barometer, in Col. 2 the simultaneous readings of the aneroid as
pressure was *reduced* inch by inch, and in Col. 3 the simultaneous readings
of the aneroid as pressure was *restored* inch by inch.

---

[1] The mercurial barometer readings, for comparison, were reduced to 32° Faht.

[2] This passage (marked by an asterisk) is a recognition of "the recovery."

1.	2.	3.
Readings of K. O. Standard Barometer.	Readings of No. 899, Cary, with pressure diminishing.	Readings of No. 899, Cary, with pressure increasing.
inches.	inches.	inches.
30·000	30·000	29·800
29·000	29·050	28·750
28·000	28·050	27·800
27·000	27·000	26·750
26·000	26·000	25·700
25·000	25·050	24·700
24·000	24·050	23·700
23·000	23·000	22·650
22·000	22·000	21·650
21·000	21·000	20·650
20·000	19·950	19·650
19·000	18·950	18·700
18·000	17·950	17·750
17·000	16·950	16·800
16·000	15·950	15·850
15·000	14·950	14·900

It will be seen that at no single inch do the aneroid readings in the two columns agree. Every reading in Col. 3 is lower than the corresponding one in Col. 2. It is apparent that in the short space of time during which this aneroid was submitted to diminished pressure (a space of time which probably did not amount to one hour in all), it showed a distinct loss upon the mercurial barometer at every inch of its scale. *The differences between the readings in the two columns at each inch represent the amounts lost at those parts of the scale during the brief time that was occupied in reducing pressure to 15 inches and restoring it again.* Thus, at 21 inches, although there is no error in Col. 2 (the aneroid reading truly against the mercurial barometer) there is a minus error of 0·350 of an inch in Col. 3. This amount (0·350 of an inch) is the amount which was lost during the time pressure was reduced inch by inch from 21 to 15 inches, and subsequently restored inch by inch to 21 inches,[1] and the same explanation applies to every other inch of the scale.

As it is demonstrable from the certificate itself that the aneroid lost upon the mercurial barometer at every inch of its scale when pressure was diminished, I am unable to see the propriety of applying as corrections the amounts which are stated on p. 51 (in the column headed "Correction to aneroid with a pressure diminishing"). For example, at 23, 22, and 21 inches it is directed that there are *no* corrections when

---

[1] It is probable that this part of the examination did not occupy so much as thirty minutes.

pressure is diminishing, yet the certificate itself shows that the aneroid lost 0·350 of an inch at each of those inches during the time it was under trial. From this it will, I think, be apparent that *it is desirable to make some radical change in the style of Certificates of Examination.*

§ 65. The last point to which I wish to draw attention is the change in index-error which occurred in this instrument. When No. 899, Cary, was examined at Kew in 1888 it had a plus error of 0·050 in. at 29 inches, when the testing was commenced (merc. bar. 29·000, aneroid 29·050). At the termination of the examination this was changed to a minus error of 0·250 in. (merc. bar. 29·000, aneroid 28·750). The aneroid recovered, and by the time it came into my hands in 1889 the error at 29 inches was + 0·073. After my testing was done, I retained the instrument for a few days to watch its recovery, and when it left me it had at 29 inches a minus error of 0·179 (merc. bar. 29·000, aneroid 28·821). It was continuing to recover, and, by the time it was first used in the field, I have no doubt it was reading closely against the mercurial barometer. But when this instrument came to me the second time (in 1890), after having experienced large variations in pressure, its index-error at 29 inches was − 0·779 of an inch (merc. bar. 29·000, aneroid 28·221). I re-tested it under the air-pump, and found that the major part of the error was carried along the scale. At 16 inches its index-error was − 0·596 of an inch (merc. bar. 16·000, aneroid 15·404). If it had been again kept at a pressure of 16 inches for a week it would again have lost a large amount, and this time the loss, so to speak, would have been piled up on the top of the index-error (− 0·596), and in the aggregate would have exceeded a barometric inch. Although the change in index-error was obvious on return to London, it would not have been so apparent to an observer in the field unless he had been possessed of the means of determining index-errors. He might, and probably would have continued to apply the corrections stated upon the Certificate of Examination.

The Kew certificate of No. 899, Cary, declares the errors which were remarked at each inch of the scale of this aneroid in Oct. 1888, upon its being reduced, inch by inch, as rapidly as possible. So long as there was no index-error at 30 inches, the corrections stated in the certificate were proper ones to employ, provided the aneroid experienced exact repetition of the same treatment. But they are not the proper ones to employ if the time conditions are different, or if there is change in the initial index-error. In the absence of directions to the contrary, it may be assumed by persons into whose hands similar certificates come that the corrections stated in them are good under all conditions and for all time ; and I have therefore thought it desirable to point out the limited value of this form of certificate, and again to dwell upon the necessity of workers in the field being able to re-determine index-errors of aneroids at any time.

§ 66. To determine the index-errors of aneroids in the field one must have recourse to the mercurial barometer—a mountain barometer, on the Fortin principle, by preference, if it can be carried. Upon some rough or prolonged journeys it is well-nigh impossible to transport mountain barometers in safety. It is, however, possible on all journeys, of whatever nature, to carry graduated (unfilled) barometer tubes and a bottle of mercury. In its simplest form, a mercurial barometer can always be available ; and by reverting to the practice of earlier travellers in employing plain or graduated tubes and filling them on the spot [1] one may combine the accuracy of the mercurial barometer with the conveniences of the aneroid.

During the progress of the experiments which are recorded in this paper, several travellers have made important journeys upon which they have relied on aneroid barometers for determination of altitude, and they have sometimes said that their aneroid observations agreed very well with their boiling-point observations. I find this difficult to understand, and I think that the difficulty will be shared by the travellers themselves if they will submit their aneroids to artificially-produced diminution of pressure for some length of time, and note the errors which they will exhibit upon the mercurial barometer.

I could not speak with any confidence upon this matter if there had been an exception to the general rule ; but I repeat that throughout the whole of my experiments in the workshop there has been no occasion upon which there has been one, and that every aneroid, without exception, which has been submitted to diminished pressure for a month, a week, or even for a day, has first lost in a marked degree upon the mercurial barometer, and has recovered upon restoration of pressure.

I invite those who are able to do so to perform such experiments as have been described in these pages, to convince themselves of the serious nature of the mistakes which must be fallen into by travellers and others if they are led to suppose that aneroids will exhibit the same errors upon the mercurial barometer at the end of a week or a month as they do in a few minutes or hours ; and also to satisfy themselves that comparisons of a traveller's aneroids against the mercurial barometer, at natural pressure, upon return to the level of the sea after prolonged journeys in elevated regions, have not the value which is at present assigned to them.

My best thanks are due to the various gentlemen who have aided me by lending aneroids,—especially to Mr. Louis Casella and to Mr. J. J. Hicks for the facilities that they have rendered to an enquiry which might have proved damaging to their interests. It is to be hoped they will have their reward. I shall have mine if the publication of this paper leads to the more scientific use of a valuable instrument.

---

[1] To be done in camp, and aneroids to be used when on the march.
Although this is a rough method, excellent results have been obtained by it in the past, and it is likely to afford a better standard for comparison than will be evolved by boiling thermometers. At the best, the boiling-water method is a circuitous manner of arriving at a result which can be obtained with greater facility and by a smaller consumption of time in another fashion. See *Travels amongst the Great Andes of the Equator*, Appendix B.

NOTE UPON THE GREENWICH TABLE OF "CORRESPONDING NUMBERS
OF ELEVATION."

The table to which I referred at p. 39 (note) is intended to be used for the graduation of Aneroids. It starts from zero (level of the sea) and gives 31 inches as the corresponding height of the Aneroid or corrected Barometer (that is the Mercurial Barometer reduced to 32° Faht.). This table is given (condensed and extended) upon the opposite page. Many instrument-makers have followed it literally, and made 31 inches their zero, or level of the sea.

At the foot of this Greenwich Table the instructions are given which are reproduced upon p. 57 ; and, if they are followed, there is no objection to 31 or even 32 inches being made zero. But in practice many persons find it impossible to follow these instructions ; many others are unacquainted with them ;[1] and yet more would not follow them if they were acquainted with them. A considerable proportion of those who employ aneroids wish to obtain a fair *notion* of their elevation above the level of the sea by mere inspection of the height indicated by the index-needle, without reference to "lower stations," or being obliged to make computations of any sort. I have frequently pointed out to instrument-makers, and think it may be useful to say here, that these persons are unnecessarily led into error through the zero being placed at 31 inches. Inasmuch as the mean Barometer at the level of the sea, in all parts of the world, is much nearer to 30 than to 31 inches,[2] the adoption of 30 inches as zero would be more appropriate. Should this be done, the "corresponding numbers" will be those given in the following table (the mean of atmospheric temperature being 50° Faht.).

Aneroid. Inches.	Height in feet.	Aneroid. Inches.	Height in feet.	Aneroid. Inches.	Height in feet.
31·000	− 894	24·500	+ 5520	18·000	+ 13,923
30·500	− 450	24·000	+ 6082	17·500	+ 14,691
30·000	0	23·500	+ 6656	17·000	+ 15,481
29·500	+ 458	23·000	+ 7242	16·500	+ 16,295
29·000	+ 924	22·500	+ 7841	16·000	+ 17,133
28·500	+ 1398	22·000	+ 8454	15·500	+ 17,998
28·000	+ 1881	21·500	+ 9081	15·000	+ 18,892
27·500	+ 2372	21·000	+ 9722	14·500	+ 19,817
27·000	+ 2872	20·500	+ 10,378	14·000	+ 20,773
26·500	+ 3381	20·000	+ 11,051	13·500	+ 21,762
26·000	+ 3900	19·500	+ 11,741	13·000	+ 22,786
25·500	+ 4429	19·000	+ 12,449	12·500	+ 23,848
25·000	+ 4969	18·500	+ 13,176	12·000	+ 24,952

[1] As they are not issued with the aneroids.
[2] See Smithsonian Miscellaneous Collections, 153 ; entitled *Tables Meteorological and Physical,* section iv (hypsometrical tables), tab. xii, p. 85.

CORRESPONDING NUMBERS OF ELEVATION IN ENGLISH FEET, AND OF READINGS
OF ANEROID IN ENGLISH INCHES (THE MEAN OF ATMOSPHERIC
TEMPERATURES BEING 50° FAHRENHEIT).

Height in feet.	Aneroid. Inches.	Height in feet.	Aneroid. Inches.	Height in feet.	Aneroid. Inches.	Height in feet.	Aneroid. Inches.
0	31·000	6000	24·875	12,000	19·959	18,000	16·016
200	30·773	6200	24·693	12,200	19·813	18,200	15·899
400	30·548	6400	24·512	12,400	19·669	18,400	15·783
600	30·325	6600	24·333	12,600	19·525	18,600	15·668
800	30·103	6800	24·155	12,800	19·382	18,800	15·553
1000	29·883	7000	23·979	13,000	19·240	19,000	15·439
1200	29·665	7200	23·803	13,200	19·100	19,200	15·326
1400	29·448	7400	23·629	13,400	18·961	19,400	15·214
1600	29·233	7600	23·457	13,600	18·882	19,600	15·103
1800	29·019	7800	23·285	13,800	18·684	19,800	14·993
2000	28·807	8000	23·115	14,000	18·548	20,000	14·883
2200	28·596	8200	22·946	14,200	18·412	20,200	14·774
2400	28·387	8400	22·778	14,400	18·277	20,400	14·666
2600	28·180	8600	22·611	14,600	18·143	20,600	14·559
2800	27·973	8800	22·446	14,800	18·011	20,800	14·453
3000	27·769	9000	22·282	15,000	17·880	21,000	14·347
3200	27·566	9200	22·119	15,200	17·749	21,200	14·242
3400	27·364	9400	21·957	15,400	17·619	21,400	14·138
3600	27·164	9600	21·797	15,600	17·490	21,600	14·035
3800	26·966	9800	21·638	15,800	17·362	21,800	13·932
4000	26·769	10,000	21·479	16,000	17·235	22,000	13·830
4200	26·573	10,200	21·322	16,200	17·109	22,200	13·729
4400	26·379	10,400	21·166	16,400	16·984	22,400	13·629
4600	26·186	10,600	21·012	16,600	16·860	22,600	13·529
4800	25·994	10,800	20·858	16,800	16·737	22,800	13·430
5000	25·804	11,000	20·706	17,000	16·615	23,000	13·332
5200	25·616	11,200	20·554	17,200	16·493	23,200	13·234
5400	25·428	11,400	20·404	17,400	16·372	23,400	13·137
5600	25·242	11,600	20·255	17,600	16·253	23,600	13·041
5800	25·058	11,800	20·107	17,800	16·134	23,800	12·946

This Table is intended more particularly for the graduation of Aneroids with a circle
of Measures in Feet concentric with the ordinary circle of Barometric Height measured in
Inches. The circle of Feet is to be read off, at the upper and lower stations, by the
Index; and the rule for measuring the height will be :—Subtract the reading at the lower
station from the reading at the upper station; the difference is the height in Feet.

The Table supposes the mean temperature of the atmosphere to be 50° Fahrenheit.
For other temperatures the following correction must be applied : Add together the
temperatures at the upper and lower station. If this sum, in degrees of Fahrenheit, is
greater than 100°, *increase* the height by $\frac{1}{1000}$ part for every degree of the excess above
100°; if the sum is less than 100°, *diminish* the height by $\frac{1}{1000}$ part for every degree of
the defect from 100°.

PART 4.—RECAPITULATION.

THE LOSS.

All aneroids lose upon the mercurial barometer when submitted to diminished pressure. §§ 1–11, 21–24, 26, 28–30, 40, 42, 51, 57–9.

When diminished pressure is maintained continuously, the loss commonly continues to augment during several weeks, and sometimes grows to a very important amount. §§ 5, 11, 21–24.

The most important part of any loss that will occur will take place in the first week. §§ 21, 25, 26.

The loss which takes place in the first week is greater than in any subsequent one. §§ 21, 23.

A considerable part of the loss which takes place in the first week occurs in the first day. §§ 28, 29.

The loss may be traced in a single hour, and in successive hours upon aneroids with expanded scales. § 30.

The amount of the loss which occurs is different in different instruments. Some lose twice, or even three times as much as others. See tables upon pp. 17, 21–24, 26.

The amount of the loss which occurs in any aneroid depends (1) upon the duration of time it may experience diminished pressure, and (2) upon the extent of the reduction in pressure. § 31.

The loss that occurs at a pressure which may be well within the range of an aneroid gives only an imperfect clue to the loss which may occur at *lower* pressures. It is commonly the case for aneroids to lose very largely indeed at the inferior limit of their range. § 31 (note).

THE RECOVERY.

When pressure is restored, all aneroids recover a portion of the loss which has previously occurred. §§ 16–18, 32–35, 37, 41, 51, 53–5, 60.

Some aneroids, in course of recovery, gain more than they have previously lost. §§ 35, 36, 37 (table). Minus index-errors are sometimes lessened, § 33 (table); plus index-errors are sometimes increased, §§ 32, 33 (table), 34 (table); and minus index-errors are sometimes converted into plus ones, § 33 (table).

The recovery is gradual, and commonly extends over a greater length of time than the period during which diminished pressure has been experienced. §§ 16, 32, 37.

In aneroids which have been kept at diminished pressures for a considerable space of time (a week and upwards), the most important part of the amount that will be recovered upon their experiencing restoration of pressure will be regained in the first week. §§ 33 (table), 35.

The greater part of the recovery of the first week is usually accomplished in the first day. §§ 34 (table), 35.

The recovery in the first hour is almost always larger than that in any subsequent hour. § 35.

TESTING.

The errors which will probably be exhibited by aneroids during natural variations of pressure may be learned approximately by submitting them to artificially-produced variations of pressure, § 19 ; but the one-hour test which has heretofore been commonly applied for 'verification' is of little value except for determining errors of graduation, or the errors which will be exhibited at similar pressures in *a similar length of time.* §§ 20, 49, 65.

It is desirable that there should be some change in the present form of Certificates of Examination, §§ 63–5 ; and that the change should be a radical one, § 64.

ON THE ANEROID AT FIXED STATIONS.

Under certain conditions, an aneroid will not follow natural diurnal or hourly variations in atmospheric pressure with reasonable accuracy. §§ 40, 41. The index of an aneroid may indicate diminishing pressure at a time when pressure is actually increasing. § 40.

ON SIMULTANEOUS COMPARISONS OF ANEROIDS.

Aneroids left at upper or lower stations (with the view of applying corrections to simultaneous observations at lower or upper stations) cannot indicate diurnal or hourly variations in pressure even with approximate accuracy unless they have remained at the level of such upper or lower stations for several days. § 42.

ON ALTERATION OF INDEX-ERRORS.

The index-errors of aneroids are seldom or never constant. § 49.

In consequence of the difference between the amounts lost and the amounts recovered, as well as through other causes, the index-errors of aneroids are liable to alter considerably. §§ 16, 17, 35–39.

A large percentage of aneroids gain under restoration of pressure a *greater* amount than they have previously lost under diminution of pressure; and a large percentage recover *less* than they have lost. It is exceptional to find the loss exactly balanced by the recovery. §§ 36, 37. In some aneroids there is a continual tendency towards the plus, and in others towards the minus direction. § 38.

Index-errors generally pass along the scale when aneroids are used rapidly. §§ 43–45, 65.

Index-errors of aneroids need to be re-determined from time to time. §§ 47–49, 62, 65.

How index-errors may be determined. § 66.

ON MEASUREMENT OF ALTITUDES.

It is probable that large reductions will have to be made in the height of many positions which have been determined by aneroid. § 27.

Reasons for this. §§ 21–26, 28–35, 40.

Ascending observations of aneroid, as a general rule [1] (provided the instruments are in "a state of repose" when starting from the lower station), indicate a greater diminution of pressure, or a greater altitude than the truth. §§ 3–9, 14, 15, 21–31, 40, 52, 57 (table), 58 (table), 59, 63 (certificate).

The error (due to loss upon the mercurial barometer) in the indicated decrease of pressure is greatest when diminished pressure (increase of elevation) has been experienced continuously for about a month, pp. 12, 13, 17 (tables); §§ 11, 21–24. In altitudes deduced from ascending observations of aneroids, errors are likely to be greatest when the instruments have been subjected continuously to diminished pressure for a month and upwards. §§ 21–26, 48.

Under certain conditions, ascending observations of aneroid may indicate less diminution of pressure, or less altitude than the truth. § 55 (note)

Provided the instruments are in "a state of repose" when starting from the upper station, descending observations of aneroid, as a general rule, indicate a greater increase of pressure, or a greater difference of level, than the truth. §§ 12, 55.

The error in the indicated increase of pressure is greatest when

---

[1] There may be exceptions to the general rule, arising from defects (such as bad graduation), or accidents which are not discussed here.

diminished pressure has been previously experienced for a month and upwards. § 55.

Under certain conditions, descending observations of aneroid may indicate less increase of pressure, or less difference of level, than the truth. §§ 57 (see table, Col. 5) ; 60, 64 (see Col. 3 of table).

Descending observations of aneroid seldom or never agree with ascending ones. § 50.

The reason of this. §§ 53–60.

To determine differences of level, observe the differences of pressure indicated on the scale of inches rather than the differences in altitude indicated on the scale of feet. § 62.

Means of ascending and descending observations are recommended in certain cases. §§ 61, 62.

Comparisons of travellers' aneroids against the mercurial barometer, at natural pressure, upon return to the level of the sea after prolonged journeys in elevated regions, have not the value which is at present assigned to them. § 17, p. 55.

THE END.

Printed by R. & R. CLARK, *Edinburgh.*

www.ingramcontent.com/pod-product-compliance
Lightning Source LLC
Chambersburg PA
CBHW022006190326
41519CB00010B/1406